Teacher's Resource Guide and Answer Key

Number Sense

Allan D. Suter

Contemporary

Series Editor: Mitch Rosin
Executive Editor: Linda Kwil
Production Manager: Genevieve Kelley
Marketing Manager: Sean Klunder
Cover Design: Steve Strauss, ¡Think! Design

Send all inquiries to:
McGraw-Hill/Contemporary
130 East Randolph Street, Suite 400
Chicago, Illinois 60601

ISBN: 0-07-287143-1

Printed in the United States of America.

3 4 5 6 7 8 9 10 QPD/QPD 09 08 07 06

The ***McGraw·Hill*** *Companies*

INTRODUCTION

To succeed in today's world, everyone needs to develop number sense. In school, at work, and in all of our daily activities, people use numbers in a variety of different ways. The needs of an increasingly global and technologically driven world require more mathematical literacy than ever before. While calculators and computers perform many needed mathematical operations, there are times when technology is not available.

The National Council of Teachers of Mathematics articulates six principles and ten standard topics for school mathematics:

PRINCIPLES

1. The Equity Principle – Excellence in mathematics education requires equity—high expectations and strong support for all students.
2. The Curriculum Principle – A curriculum is more than a collection of activities: it must be coherent, focused on important mathematics, and well articulated across the grades.
3. The Teaching Principle – Effective teaching requires understanding what students know and need to learn and then challenging and supporting them to learn it well.
4. The Learning Principle – Students must learn mathematics with understanding, actively building new knowledge from experience and prior knowledge.
5. The Assessment Principle – Assessment should support the learning of important mathematics and furnish useful information to both teachers and students.
6. The Technology Principle – Technology is essential in teaching and learning mathematics; it influences mathematics that is taught and enhances students' learning.

STANDARD TOPICS

1. Number and Operations
2. Algebra
3. Geometry
4. Measurement
5. Data Analysis and Probability
6. Problem Solving
7. Reasoning and Proof
8. Communication
9. Connections
10. Representation

With these principles and standards in mind, math educators are revising math curriculum and developing innovative methods of delivering mathematics instruction. McGraw-Hill/ Contemporary's *Number Sense* series has been revised according to these principles and standards. It is targeted to students who respond well to an activity-based, visual approach. This includes students in all areas of math education — particularly those who need extra review of basic mathematic concepts.

The *Teacher's Resource Guide and Answer Key* provides a variety of supplemental activities that are designed to introduce and/or reinforce new concepts. Many of these provide an opportunity for the students to experience mathematical concepts through hands-on learning. They can provide additional learning opportunities for students and the chance to explore a concept in greater depth.

Number Sense helps students:

- bridge the gap between the concrete and the symbolic by relating models, pictures, and diagrams to mathematical ideas
- discover basic math concepts and relationships
- acquire the language of mathematics
- create and solve problems, with a special focus on real-life problems
- use a success-oriented approach that builds confidence and independence

The *Number Sense* series promotes a high level of student involvement based on the satisfaction of task completion. One-page lessons, contained in manageable books, encourage students to feel that they are mastering concepts and making progress. The integration of computation, number relationships, word problems, life-skills tasks, and section reviews provides a basis for students to apply math concepts in a variety of ways.

Overall, McGraw-Hill/Contemporary's *Number Sense* series promotes the idea that math can be mastered, and that it can be a rewarding experience.

Using McGraw-Hill/Contemporary's *Number Sense* System

McGraw-Hill/Contemporary's *Number Sense* system consists of:

- Series Diagnostic & Placement Test for the *Number Sense* series
- Ten student books

  Whole Numbers: Addition & Subtraction

  Whole Numbers: Multiplication & Division

  Decimals: Addition & Subtraction

  Decimals: Multiplication & Division

  Fractions: The Meaning of Fractions

  Fractions: Addition & Subtraction

  Fractions: Multiplication & Division

  Ratio & Proportion

  The Meaning of Percent

  Percent Applications
- Diagnostic/Placement Pretests and Mastery Posttests in each book
- The Teacher's Resource Guide and Answer Key

The Series Diagnostic & Placement Test offers teachers the ability to assess a student's competency in all mathematic areas covered by the *Number Sense* series. The test correlations allow teachers to identify specific areas of weakness and to assign specific books for students to begin their study of mathematics.

The *Number Sense* student books offer a unique visual and activity-oriented approach that is complemented by classroom activities in the *Teacher's Resource Guide and Answer Key*. These include games, manipulatives, group activities, reinforcement tasks, and language of math activities.

Placement Pretests and the Mastery Posttests for each book in the series are powerful components of the program. Instructors can administer any of these before a student begins work and after lessons have been taught. If, on a Placement Pretest, a student shows skill deficiencies in a certain area, it is best if the student starts at the beginning of the corresponding book. The books are carefully structured so that ideas build upon one another, allowing the student to genuinely build a strong foundation in the basic steps. It is important to note, however, that each book can stand alone, and the series does not need to be taught in any particular order.

If you wish to customize a program to provide remediation of skills for a student, you can use the management system outlined on the next page.

Diagnose

1. Administer the Series Diagnostic & Placement Test.
2. Use the Evaluation Chart in the *Teacher's Resource Guide and Answer Key* to identify specific skill deficiencies.

Prescribe

1. Prescribe books as indicated by the Evaluation Chart.
2. Use the Pretest in the front of the student book to assess specific skills in the prescribed book.
3. After student completes prescribed pages, use the section review pages in student texts to monitor mastery.

Assess

1. Administer the Mastery Posttest.
2. Use the Evaluation Chart to identify further areas of weakness.
3. Reassign specific units to reinforce weak skills.
4. Use materials and games from *Teacher's Resource Guide and Answer Key* to increase student understanding.

THE LANGUAGE OF MATHEMATICS

The language of math generally uses four categories of symbols. You can introduce these symbols as students move through the different levels of understanding.

1. Ideas – numbers represent the ideas
2. Operation Symbols – what is being done with the ideas: +, −, x, ÷
3. Number Relation Symbols – how the ideas are related to one another: =, ≠, >, <, etc.
4. Punctuation Symbols – the decimal point, comma, percent sign, parenthesis, brackets, braces, etc. Many of the punctuation symbols indicate the order in which the problem is to be completed.

Additionally, many lessons in the *Teacher's Resource Guide and Answer Key* are marked with the symbol shown on the left. This symbol indicates that information has been provided for instruction in the language of math. These sections provide teachers with activities that focus on the language used in mathematics, and can be used with ESL students or students requiring additional review of math terms.

THE FOUR BASIC LEVELS OF MATHEMATICS UNDERSTANDING

The more that calculators and computers are used in our society, the greater will be the need for a solid foundation of number sense. Students must be confident in their abilities to communicate, reason, estimate, and do a good job with the things that calculators and computers cannot do well.

How well students solve problems is determined by the methods and strategies that are employed. Regardless of what types of information and instructional approaches are used, students need to move through the four basic levels of understanding.

1 Concrete Level

The first level of understanding is the concrete level. This is the level on which we are appealing to the senses, and the student is able to see, touch, handle, and move objects. For example, if I am holding $5 in one hand and $10 in the other, I have $15.

SAMPLE LESSON

Arrange clothespins on a hanger. Let students manipulate the clothespins to represent various mathematical operations and become familiar with the different phrases associated with mathematics:

2 clothespins plus 3 clothespins equals how many clothespins?

2 clothespins and 3 clothespins are how many altogether?

2 clothespins and 3 clothespins total what number?

2 clothespins and 3 clothespins equal how many clothespins?

To assist in the development of math language, have students write problems using the manipulatives. These problems can be mathematical number sentences or written word problems. Then students can exchange their problems and solve those of their classmates.

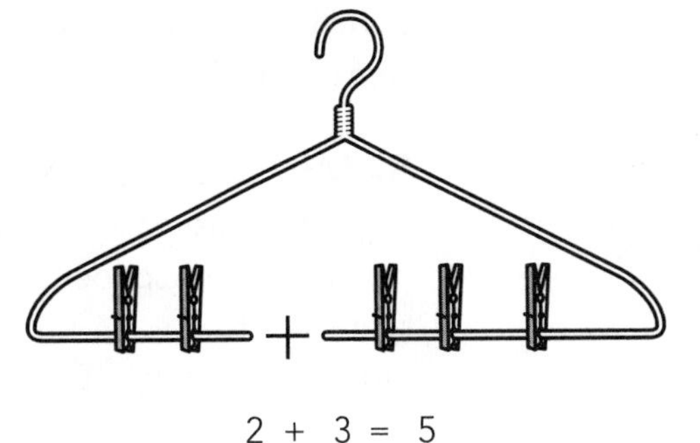

2 Semiconcrete Level

The second level of understanding is the semiconcrete level. This level is sometimes combined with level three to form the representational level. This is the level where students move from having real objects in front of them, to having pictorial representations of the real objects. There must be a transfer from what is physically present (concrete) to what is being represented in a drawing (semiconcrete). The picture-to-word strategy (semiconcrete) helps students relate concepts to real life.

SAMPLE LESSON

Present students with pictured sequences as shown below and have them orally respond to form a problem.

Paper $.08

Pencil $.15

The paper and pencil cost how much altogether?

The pencil costs how much more than the paper?

$.08 increased by $.15 is how much?

$.08 combined with $.15 equals what number?

The cost of the paper and pencil equals how much?

This level can easily be combined with level 3 to make one lesson.

3 Semiabstract Level

The third level of understanding is the semiabstract level. This level is sometimes combined with level two to form the representational level. As students move to symbolic representation, they are beginning to enter the abstract level. For example, the symbol 5 is used to represent 5 dollars, 5 pencils, or 5 cars. The symbol 5 is used no matter what objects are being represented. This is an abstraction.

SAMPLE LESSON

Have students make up their own number sentences using pictures from magazines, newspapers, or any printed source. They should write number sentences to
organize their thinking and to represent their pictorial representations. By answering problems, students will clarify their learning of the math process. This is a good time to introduce simple word problems and to teach students to ask themselves, "Does this answer make sense?" This question is very important to the successful completion of math problems.

Joe has $3. Mary has $2 more than Joe. How much money does Mary have?

$3 + $2 = $5

This is the level to introduce more language of math. Students often have a good listening and speaking vocabulary that exceeds their reading or writing ability. Have students ask each other math problems and write down the equations. They can solve the problem in their minds, without having to read the actual problem. By verbalizing and mentally solving math problems, students are reinforcing the concepts they have learned.

4 Abstract Level

The fourth level of understanding is the abstract level. This is the level where students can apply the symbols of mathematics to various problems without needing to use pictorial representations.

SAMPLE LESSON

It is never too early to work with the abstract level by introducing the algebraic concept of the unknown. You can work with unknowns in number sentences in a variety of ways related to word problems.

Rachel needed 40 tulip bulbs. She only had 17. How many more did she need?

$40 - x = 17$ OR $40 - 17 = x$

Patterns and variables can also be introduced.

List all possible combinations that equal 8.

$\square + \square = 8$

1 + 7   3 + 5

2 + 6   4 + 4

By working with all four levels of understanding, students will be able to better understand mathematical ideas so they can remember and apply concepts in the future.

CASAS Correlations

WHOLE NUMBERS: ADDITION AND SUBTRACTION

Section	Competency Number	Description
All	6.1.1	Add whole numbers
All	6.1.2	Subtract whole numbers

WHOLE NUMBERS: MULTIPLICATION AND DIVISION

Section	Competency Number	Description
Pgs. 70–74	1.2.4	Compute unit pricing
Pgs. 70–74	4.2.1	Interpret wages, deductions, benefits
Pgs. 7–29, 62–74	6.1.3	Multiply whole numbers
Pgs. 30–74	6.1.4	Divide whole numbers
Pgs. 30–74	6.9.1	Estimate answers

DECIMALS: ADDITION AND SUBTRACTION

Section	Competency Number	Description
Pgs. 36–71	1.1.6	Count, convert and use coins and currency
Pg. 71	1.2.4	Compute unit pricing
All	6.2.1	Add decimal fractions
All	6.2.2	Subtract decimal fractions
Pgs. 72–73	6.8.1	Interpret statistical information used in news reports

DECIMALS: MULTIPLICATION AND DIVISION

Section	Competency Number	Description
Pgs. 50–51	1.1.8	Compute averages
Pgs. 24–26, 69, 70	1.2.2	Comparison shop: price, quality
Pgs. 68–73	1.2.4	Compute unit pricing
Pgs. 72–73	2.6.4	Interpret restaurant menus and compute costs
Pgs. 66–67	4.2.1	Interpret wages, deductions, benefits
Pgs. 71–73	5.4.2	Compute or define sales tax
Pgs. 18–23	6.1.5	Perform multiple operations using whole numbers
All	6.2.3	Multiply decimal fractions
All	6.2.4	Divide decimal fractions
All	6.2.5	Perform multiple operations using decimal fractions

FRACTIONS: MEANING OF FRACTIONS

Section	Competency Number	Description
Pgs. 64–67	1.1.3	Interpret maps and graphs
Pgs. 68–73	1.1.6	Count, convert and use coins and currency
All	6.3.4	Divide common or mixed fractions
All	6.3.6	Convert common or mixed fractions to decimals
All	6.3.7	Identify or calculate equivalent fractions

FRACTIONS: ADDITION AND SUBTRACTION

Section	Competency Number	Description
Pgs. 68–69	1.1.1	Interpret recipes
All	6.3.1	Add common or mixed fractions
All	6.3.2	Subtract common or mixed fractions
All	6.3.5	Perform multiple operations using fractions
Pgs. 68–71	6.6.4	Use or interpret measurement instruments such as rulers, scales, gauges, and dials

FRACTIONS: MULTIPLICATION AND DIVISION

Section	Competency Number	Description
Pgs. 70–71	1.1.1	Interpret recipes
All	6.3.3	Multiply common or mixed fractions
All	6.3.4	Divide common or mixed fractions
All	6.3.4	Perform multiple operations using common or mixed fractions

RATIO AND PROPORTION

Section	Competency Number	Description
Pgs. 53–55	6.2.1	Convert decimal fractions to common fractions
Pgs. 34–67	6.3.4	Divide common or mixed fractions
All	6.4.6	Compute using ratio or proportion

MEANING OF PERCENT

Section	Competency Number	Description
All	6.4.2	Apply a percent in a context not involving money
All	6.4.3	Calculate percents
All	6.4.4	Convert percents to fractions or decimals
All	6.4.6	Compute using ratio or proportion

PERCENT APPLICATIONS

Section	Competency Number	Description
Pgs. 59–61	1.2.2	Comparison shop: price, quality
Pgs. 37–61	1.2.4	Compute unit pricing
Pg. 65	1.8.3	Interpret interest or interest-earning savings
Pgs. 62–64	5.4.2	Compute or define sales tax
Pg. 28	6.3.6	Convert common or mixed fractions to decimals
Pgs. 7–12, 57–74	6.4.1	Apply a percent to determine amount of discount
Pgs. 7–12, 57–74	6.4.2	Apply a percent in a context not involving money
All	6.4.3	Calculate percent
Pgs. 14–16	6.4.4	Convert percents to common, mixed, or decimal fractions
Pg. 66	6.5.1	Recognize and evaluate simple consumer formulas

Solve each problem.

1. $\begin{array}{r} 93 \\ +78 \\ \hline \end{array}$

Answer: ________

2. $54 + 376 + 89 =$

Answer: ________

3. $\begin{array}{r} 413 \\ -258 \\ \hline \end{array}$

Answer: ________

4. $6{,}034 - 1{,}967 =$

Answer: ________

5. Before lunch on Monday, Fred drove 209 miles. After lunch, he drove another 263 miles. How far did he drive altogether that day?

Answer: ________

6. Toby's take-home pay rose from $583 a week to $619 a week. By how much did her pay increase?

Answer: ________

7. $\begin{array}{r} 479 \\ -\quad 6 \\ \hline \end{array}$

Answer: ________

8. $902 - 48 =$

Answer: ________

9. $7\overline{)511}$

Answer: ________

10. $7{,}542 \div 18 =$

Answer: ________

11. Mary Lou pays $560 a month for rent for her apartment. How much rent does she pay in one year?

Answer: ________

12. Tara makes $595 for a 35-hour work week. How much does she make per hour?

Answer: ________

13. 9.26 + 23 =

Answer: ________

14. 4.8 + 27.09 + 6.763 =

Answer: ________

15. 7.2 − 1.83 =

Answer: ________

16. 24 − 11.9 =

Answer: ________

17. Ted wants to mail two packages that weigh 4.3 pounds and 11.28 pounds. What is the combined weight of the packages?

Answer: ________

18. The car's mileage gauge read 36.4 when it left the dealership, and 408.7 at the end of the first week. How many miles were driven that first week?

Answer: ________

19.
$$\begin{array}{r} 5.13 \\ -\ .9 \\ \hline \end{array}$$

Answer: ________

20. .27 − 6.8 =

Answer: ________

21. $.8\overline{)6.96}$

Answer: ________

22. $72 \div .12 =$

Answer: ________

23. Find the total weight of five boxes if each box weighs 4.6 pounds.

Answer: ________

24. Dorothy road her bicycle 117.6 miles from Burnaby to Whistler in 6 hours. Find her average speed in miles per hour.

Answer: ________

25. Change $\frac{35}{50}$ to a fraction in simplest form.

Answer: ________

26. What is $\frac{24}{36}$ in simplest form?

Answer: ________

27. Change $\frac{3}{4}$ to an equivalent fraction with a denominator of 16.

Answer: ________

28. Write $\frac{5}{8}$ as a fraction with a denominator of 40.

Answer: ________

29. What is the decimal form of the improper fraction $\frac{9}{4}$?

Answer: ________

30. On a test with 20 questions, Bobbi got 16 questions correct. What fraction of the questions did she get right? Express your answer in simplest form.

Answer: ________

31. $\begin{array}{r} 3\frac{5}{8} \\ +4\frac{7}{8} \\ \hline \end{array}$

Answer: ________

32. $5\frac{1}{2} + 2\frac{3}{10} + 3\frac{1}{3} =$

Answer: ________

33. $\begin{array}{r} 7\ \frac{5}{9} \\ -\ 4\ \frac{1}{3} \\ \hline \end{array}$

Answer: ________

34. $8\frac{1}{4} - 5\frac{3}{5} =$

Answer: ________

35. Joyce's table is $59\frac{1}{2}$ inches long. With an extension panel, the table becomes $15\frac{3}{4}$ inches longer. What is the length of the table with the extension?

Answer: ________

36. From an 18-inch length of pipe, Mereille cut a piece $11\frac{9}{16}$ inches long. How long was the remaining piece of pipe?

Answer: ________

37. $\frac{3}{4} - \frac{2}{5} =$

Answer: ________

38. $1\frac{4}{5} - 2\frac{2}{3} =$

Answer: ________

39. $3\frac{1}{2} \div \frac{3}{4} =$

Answer: ________

40. $\frac{5}{12} \div 2\frac{1}{2} =$

Answer: ________

41. A soup recipe calls for $\frac{2}{3}$ cup of beans. For a big family dinner, Henrik will make four times the recipe. How many cups of beans does he need for the soup?

Answer: ________

42. How many wooden brackets each $1\frac{1}{4}$ feet long can be cut from a 10-foot board?

Answer: ________

43. Write the ratio of 12 to 32 in simplest form.

Answer: ________

44. Simply the ratio 50:60.

Answer: ________

45. Solve for x in $\frac{x}{8} = \frac{7}{20}$.

Answer: ________

46. Solve for n in $5:24 = 9:n$.

Answer: ________

47. Ten of the 24 employees where Adrian works did not vote. What is the ratio of the employees who voted to the total number of employees?

Answer: ________

48. Chris can drive his truck 35 miles on 2 gallons of gasoline. How many gallons of gasoline does Chris need to drive 175 miles?

Answer: ________

49. Change $\frac{9}{20}$ to a percent.

Answer: ________

50. What is 28% as a fraction in simplest form?

Answer: ________

51. Write the whole number 6 as a percent.

Answer: ________

56. Find 18% of \$630.

Answer: ________

52. Change 0.145 to a percent.

Answer: ________

57. 32 is what percent of 40?

Answer: ________

53. Write 7.2% as a decimal.

Answer: ________

58. What percent of 96 is 12?

Answer: ________

54. Change 2.4 to a percent.

Answer: ________

59. A \$189 lawn mower was on sale for 15% off. How much can John save if he purchases the lawn mower on sale?

Answer: ________

55. What is 25% of 136?

Answer: ________

60. Lin recorded the rainfall on 18 days in June. For what percent of the days in June was rain recorded?

Answer: ________

Placement Test Answer Key

1. 171
2. 519
3. 155
4. 4,067
5. 472 miles
6. $36
7. 2,874
8. 43,296
9. 73
10. 419
11. $6,720
12. $17
13. 32.26
14. 38.653
15. 5.37
16. 12.1
17. 15.58 pounds
18. 372.3 miles
19. 4.617
20. 1.836
21. 8.7
22. 600
23. 23 pounds
24. 19.6 mph
25. $\frac{7}{10}$
26. $\frac{2}{3}$
27. $\frac{12}{16}$
28. $\frac{25}{40}$
29. 2.25
30. $\frac{4}{5}$
31. $8\frac{1}{2}$
32. $11\frac{2}{15}$
33. $3\frac{2}{9}$
34. $7\frac{13}{20}$
35. $75\frac{1}{4}$ inches
36. $6\frac{15}{16}$ inches
37. $\frac{3}{10}$
38. $4\frac{4}{5}$
39. $4\frac{2}{3}$
40. $\frac{1}{6}$
41. $2\frac{2}{3}$ cups
42. 8 brackets
43. 3:8
44. 10:13
45. $2\frac{4}{5}$
46. $43\frac{1}{5}$
47. 7:12
48. 10 gallons
49. 45%
50. $\frac{7}{25}$
51. 600%
52. 14.5%
53. 0.072 or .072
54. 240%
55. 34
56. $113.40
57. 80%
58. $12\frac{1}{2}$% or 12.5%
59. $28.35
60. 60%

Placement Test Evaluation Chart

Problem Numbers	Number Sense Book
1, 2, 3, 4, 5, 6	Whole Numbers: Addition & Subtraction
7, 8, 9, 10, 11, 12	Whole Numbers: Multiplication & Division
13, 14, 15, 16, 17, 18	Decimals: Addition & Subtraction
19, 20, 21, 22, 23, 24	Decimals: Multiplication & Division
25, 26, 27, 28, 29, 30	Fractions: The Meaning of Fractions
31, 32, 33, 34, 35, 36	Fractions: Addition & Subtraction
37, 38, 39, 40, 41, 42	Fractions: Multiplication & Division
43, 44, 45, 46, 47, 48	Ratio & Proportion
49, 50, 51, 52, 53, 54	The Meaning of Percent
55, 56, 57, 58, 59, 60	Percent Applications

WHOLE NUMBERS
Addition & Subtraction

Our number system is called the decimal system because it is based on the number 10. The Latin prefix *deci* means tenth. It is important that students understand the basic structure of the whole-number system in preparation for the basic operations.

Before teaching addition and subtraction of whole numbers, it is helpful for students to experience identifying, observing, and visualizing the concepts of place value. Once students understand the decimal system, it is easier for them to learn the principles of regrouping, also known as carrying and borrowing.

Student Glossary

Acquainting students with definitions of key math terms and life-skills concepts will enhance their mastery of the materials. Below are words defined in the student text. A glossary is provided at the end of the student text and on page 243 of the *Teacher's Resource Guide and Answer Key.*

addition
attendance
check
checking account
digit
difference
expanded form
general admission
minus
number relation symbol
number sentence
operation symbol
opposite
options
pattern
place value
plus
quoted
registered
regroup (borrow)
regroup (carry)
regular price
reserved seats
sale price
subtraction
sum
symbol
take away

STUDENT PAGE
7
Place Value

50 and Out (Place Value Game)

Having a visual representation of place value helps students to conceptualize place value.

1 Divide a playing board into two halves. Write *ones* on one side and *tens* on the other.

2 Place a pile of straws on the *ones* side.

3 Have players take turns throwing one die and picking up straws for the number that appears on the die.

4 As soon as a student has ten straws, he or she can put a rubber band around them and place them on the tens side. The first player with 50 straws, or 5 bundles, wins the game.

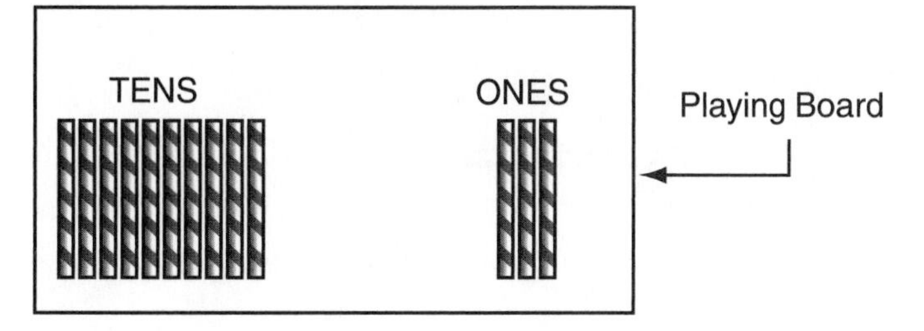

STUDENT PAGE
7
Place Value

Place Value and Money

Using dimes and pennies is a good way to represent two-digit numbers. If possible, demonstrate with real or play money. Use the least number of coins possible.

EXAMPLES

79 cents or 7 dimes + 9 cents or seventy-nine cents

28 cents or 2 dimes + 8 cents or twenty-eight cents

10 cents or 1 dime + 0 cents or ten cents

Alternate Activity

Using only dimes and pennies, show how you would pay for:

a 25¢ candy bar ______________ + ______________
dimes pennies

a 68¢ pen ______________ + ______________
dimes pennies

a 92¢ ice cream cone ______________ + ______________
dimes pennies

STUDENT PAGE
9
Place Value

Newspaper Math

Cut out the numbers used in a newspaper story and pass them around the class. Ask the students to fit the numbers back into the sentences in their correct positions. This activity gives students a sense of the importance of numbers.

EXAMPLE

Each school currently pays ______ dollars to send ______ students on a trip to the state capital. This covers only ______ of their operating costs.

one-third 30 300

STUDENT PAGE
10
Place Value

Use Real-Life Numbers

Newspapers and magazines are an everyday illustration of mathematics. Nearly every article and advertisement contains numbers. You can invent and expand on ideas from newspapers and magazines.

EXAMPLE

Large numbers are common in today's world of trillion-dollar national budgets. Choose articles that discuss large sums of money. Have your students read the numbers as words, reviewing the place-value structure of our number system. Discuss why it is advantageous to round numbers when communicating with the public. Have students write numbers that are presented in word form.

STUDENT PAGE
13
Place Value

Discussing Place Values

You can lead a discussion of place values by writing problems like these on the board.

893 = 800 + ______ + 3 Ask students to explain what the 9 in 893 stands for.

460 = ______ + 60 + 0

2,974 = ______ + ______ + ______ + ______

6,005 = ______ + ______ + ______ + ______ Discuss zeros as placeholders.

600 + 20 + 8 = ______

400 + 9 = ______

3,000 + 500 + 30 + 6 = ______

STUDENT PAGE
14
Place Value

Rearrange the Digits

On the board, write a number with two or more digits. For example: 6,931. Rearrange the digits to name the smallest, then the largest, possible number. Have students read the number after it has been rearranged. Students can also rearrange digits for other students to read.

STUDENT PAGE
15
Place Value

Guess the Number

Think of a number less than 100. Keep track of how many guesses it takes for students or groups of students to reach the number. The game continues until the number is picked.

EXAMPLE

Choose the number 25. Have the students guess.

Guess 1: 56 **Response:** The number is less than 56.

Guess 2: 20 **Response:** The number is greater than 20.

Guess 3: 40 **Response:** The number is less than 40.

Guess 4: 30 **Response:** The number is less than 30.

Alternate Activity

Draw a number line on the board with reference points.

EXAMPLE

0 and 100
Think of a number less than 100. For example: 65.

0 30 75 100

Guess 1: 30 Have a student come to the board and mark 30 on the line. Tell them that the number is greater than 30.

Guess 2: 75 Ask another student to come to the board and mark 75 on this line. Tell them that the number is less than 75.

The game continues until the number is chosen.

STUDENT PAGE
15
Place Value

Estimating the Size of Numbers

From newspaper, magazines, encyclopedias, almanacs, etc., assign students to find numbers of different values. Scramble the numbers in a box. Challenge the class to match the scrambled numbers to a sentence. This helps students get a realistic sense of the relative sizes of numbers.

EXAMPLES

The height of Mt. Everest is ____________ feet.

The sequoia (big tree) named German Sherman in Sequoia National Park, California, is almost ____________ feet tall.

30	91	275	15	5,200,000	29,000

Some pigeons have been recorded to fly as fast as ____________ mph.

If your weight is 80 pounds, about ____________ pounds is muscle.

About ____________ cars are sold each year.

STUDENT PAGE 18 Addition

Find the Numbers

Tell students that the □ and △, as shown below, represent different numbers. Have students make a chart to represent addition facts. This nontraditional method will help students learn the facts as a puzzle, not as rote memorization.

EXAMPLES

List all possible combinations that equal 7.

□	△
5	2
6	1

$\square + \triangle = 7$

etc.

List all possible combinations that equal 10.

□	△
8	2
7	3

$\square + \triangle = 10$

etc.

STUDENT PAGE 18 Addition

Review the Facts

To review basic addition facts, use a scrambled grid. You may add more numbers along the top and side. As students become more comfortable with whole numbers, they can create their own grids and solve each other's.

EXAMPLE

+	5	8	7	7
3	8	11		
6				
2			9	
5				

STUDENT PAGE 19 Addition

Addition Speed Drill

Use a timer on addition facts only after the students have had concrete experiences and have demonstrated an understanding of the operation. Most students enjoy flash cards and timed fact tests.

STUDENT PAGE 19 Addition

Find the Sum

List 4 numbers on the board and have students mentally complete number sentences. The degree of difficulty can be varied by changing the size of the numbers.

EXAMPLE

[7] [9] [5] [2]

16 = ________ + ________

11 = ________ + ________

7 = ________ + ________

9 = ________ + ________

12 = ________ + ________

14 = ________ + ________

Alternate Activity

List 4 numbers on the board and have your students write 6 different addition number sentences based on the numerals. Challenge students to write as many number sentences as they can from the numbers.

STUDENT PAGE 20 Addition

Skip Counting

A tennis ball is tossed back and forth in a group of students. The student catching the ball must name the next multiple. The multiple can be called out while the ball is in the air or after it is caught.

EXAMPLES

4, 8, 12, 16 . . .

5, 10, 15, 20 . . .

6, 12, 18, 24 . . .

The team with the highest multiple at the end of one minute wins.

STUDENT PAGE
20
Addition

Search for a Pattern

Have students extend patterns. Sometimes multiples can be used, and other times the student must search for a pattern.

EXAMPLES

8, 16, 24, ______, ______, ______, ______, ______, ______, 80

5, 12, 19, ______, ______, ______, ______, ______, ______, 68

10, ______, 30, ______, 50, ______, ______, ______, ______, 100

Let students make up their own patterns and share them with the class. The degree of difficulty will vary according to the ability of your students.

STUDENT PAGE
21
Addition

Creating Stories

Every student should be able to think about a mathematical statement as a story about reality. Basic number facts in the horizontal form are called **number sentences.** Ask students to make up their own stories from number sentences written on the board.

EXAMPLE

9 + 5 = 14; $5 added to the $9 already saved is ____________.

5 more than 9 is ____________.

9 marbles and 5 marbles combined equal ____________.

9 and 5 equal ____________.

9 people were on the bus. 5 more got on. There were ____________ people on the bus.

STUDENT PAGE
22
Addition

Column Addition—Adding Three or More Numbers

You can use pictorial models or concrete manipulatives to introduce column addition to students. For example, have students group the manipulatives together in the same way as they're being taught to group numbers.

EXAMPLE

$$\begin{array}{r} 3 \\ 2 \\ +\ 4 \\ \hline 9 \end{array} \rightarrow \begin{array}{r} 5 \\ \\ 4 \\ \hline 9 \end{array}$$

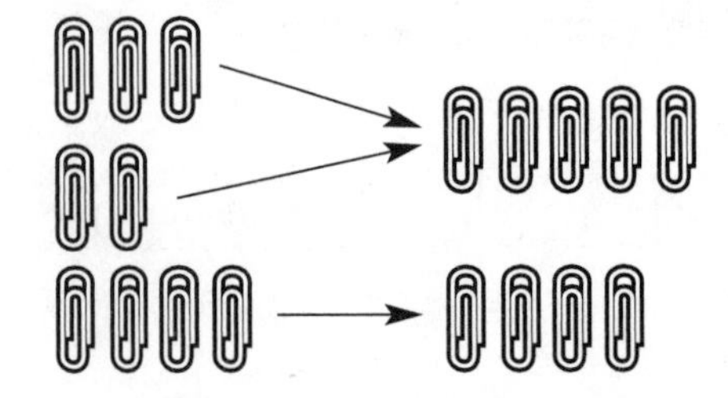

Alternate Activity

Number lines also work well to demonstrate column addition.

EXAMPLE

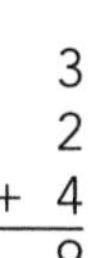

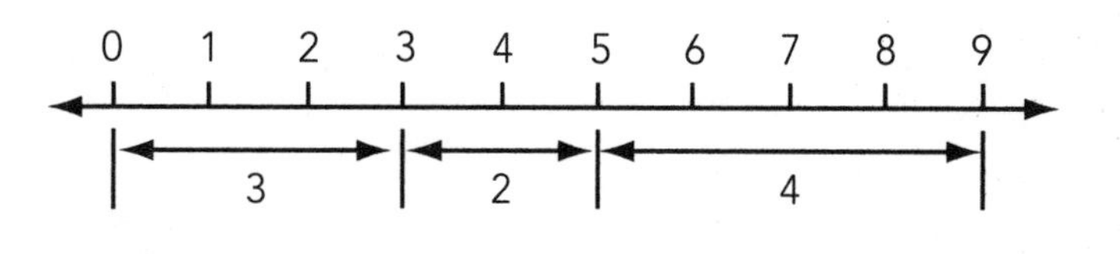

STUDENT PAGE **23** Addition

Real-Life Addition

To reinforce addition concepts, discuss real-life problems that involve money.

EXAMPLE

Sue had $36 in the bank and made a deposit of $52. How much money does she have in the bank after making the deposit?

36	=	3 ten-dollar bills + 6 one-dollar bills
+52	=	5 ten-dollar bills + 2 one-dollar bills
88	=	8 ten-dollar bills + 8 one-dollar bills
	=	80 dollars + 8 dollars
	=	88 dollars

STUDENT PAGE **23** Addition

Mentally Find the Sum

Scramble a series of numbers on the blackboard and have your students list as many pairs whose sum is 100 as they can in one minute. This can also be done for other sums; for instance, 50, 75, etc.

EXAMPLE

Find 100

36 80 20 64 60 25 10 40 75 90

STUDENT PAGE **25** Addition

Guess and Check

Have students complete number sentences. In each number sentence, the boxes must represent the same number. So, whatever number they write in one box must be written in the others. Do several of these with students so they understand the concept. Have them check their answers by substituting their answers for the boxes.

EXAMPLES

A ☐ + ☐ + 4 = 22 (☐ = 9)

B ☐ + ☐ + 8 = 7 + 5 (☐ = 2)

C ☐ + 15 + ☐ = 31 (☐ = 8)

D ☐ + ☐ + ☐ = 6 + 15 (☐ = 7)

STUDENT PAGE **26** Addition

Regrouping Practice

Regrouping, or **carrying**, requires a good understanding of place value. To reinforce regrouping, have students work examples using expanded place-value ideas.

EXAMPLE 1

65 = 6 tens + 5 ones
+ 9 = 0 tens + 9 ones
= 6 tens + 14 ones
= 7 tens + 4 ones
= 70 + 4
= 74

EXAMPLE 2

67 = 6 tens + 7 ones
+ 75 = 7 tens + 5 ones
= 13 tens + 12 ones
= 14 tens + 2 ones
= 140 + 2
= 142

STUDENT PAGE **26** Addition

Palindromic Numbers

A **palindromic number** reads the same backward and forward.
For example, 363 or 838
Any number can be made into a palindromic number.

EXAMPLE

58	Starting number
+ 85	Reverse the digits and add.
143	Sum is not a palindromic number, so
+ 341	reverse the digits and add again.
484	The sum is a palindromic number after 2 reversals.

For any number, students can add the previous sum and its reverse, and it will eventually produce a palindromic number. Students enjoy this activity, which provides addition practice in a nontraditional format. There are some numbers that will take more than ten reversals to produce a palindromic number. Students can keep a chart to record their results.

EXAMPLES

Number	Reversals	Palindromic Number
58	2	484
374	5	44044

STUDENT PAGE 28 Addition

Biggest Sum Wins

1 Present the class with an addition format similar to one of the formats below.

2 From a set of cards marked 0–9, each player or team takes a turn choosing one random card at a time.

3 Students place the digit in one of the boxes of the format and return the card to be mixed in for the next draw.

4 The object of the game is to build numbers that can be added to result in the largest possible sum. The level of difficulty can be adjusted by using different formats.

EXAMPLE FORMATS

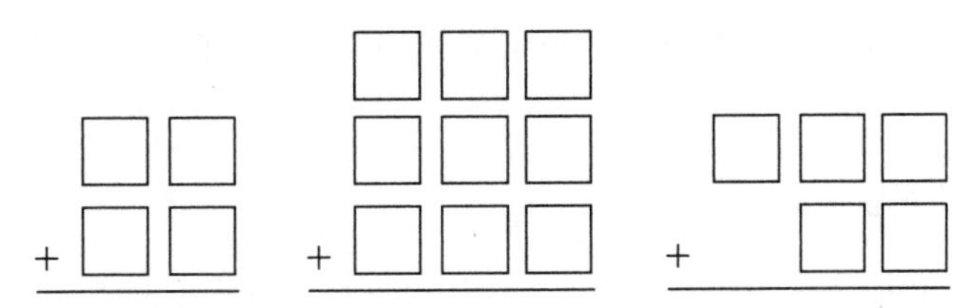

Alternate Activity

The player or team with the smallest answer wins.

STUDENT PAGE 31 Addition

Estimation Strategies

Estimation activities should be presented not as separate lessons but as a skill used in all computational activities.

EXAMPLES

A Comparison

48 + 36 Both numbers are less than 50, so the total must be less than 100.

58 + 62 Both numbers are greater than 50, so the total must be more than 100.

B **Clustering**

66 + 28 + 74 + 34 Two of the numbers are close to 70, and two are close to 30. The sum can be estimated as: (70 + 30) + (70 + 30) = 200.

128 + 55 + 48 + 88 Two of the numbers are close to 100, and two are close to 50. The sum can be estimated as: (100 + 50) + (50 + 100) = 300.

STUDENT PAGE **31** Addition

Estimate the Answer

Make flash cards with estimation exercises and answer choices. This provides an excellent speed drill. There are many ways to make estimates. Discuss with your students how they arrived at their estimates.

EXAMPLES

Problems	Estimated Answers		
321 + 593 =	500	900	1,000
78 + 85 =	100	200	300
879 + 115 =	900	1,000	1,100

STUDENT PAGE **31** Addition

Mental Arithmetic

Practicing mental arithmetic should be a regular part of an instructional program.

EXAMPLES

A Show students how to compute mentally from left to right:

634 + 338 = ______________

Add the hundreds, tens, and ones separately and then add the sums (900 + 60 + 12). Problems such as these should be done without using pencil and paper.

B 65 + 39 = ______________

To eliminate the need to carry, the student adds 65 and 40 and subtracts 1 to get 104.

STUDENT PAGE **32** Addition

Fill In the Missing Digits

This type of problem offers a good review of adding whole numbers.

EXAMPLES

$$\begin{array}{r} 8\square \\ +\ 9\ 3 \\ \hline 1\square 2 \end{array} \qquad \begin{array}{r} 3\ 5\ 8 \\ +\ \ \square 3 \\ \hline \square 3\ 1 \end{array} \qquad \begin{array}{r} 2\ 3\ 5 \\ 5\ \square\ 7 \\ +\ \square\ 4\ 2 \\ \hline 1\ 7\ 5\ \square \end{array}$$

STUDENT PAGE
35
Addition
Problem Solving

From Word Problems to Number Sentences

Word problems can be presented verbally by a student or teacher, recorded on a tape recorder, or written on cards. Have students listen to and/or read a problem and then write a number sentence based on it. This activity helps students acquire the language of mathematics and become active participants in creating and solving problems.

STUDENT PAGE
39
Subtraction

Subtraction Models

Make pictorial models or use manipulatives such as paper clips or poker chips to show "take away".

EXAMPLE

Ask students, "If we have 5 paper clips and 'take away' 2, how many are left?"

5 – 2 = 3

Alternate Activity

A → ○○○○○ Set A has 5 circles.

B → ○○ Set B has 2 circles.

How many more circles are there in set A?
How many fewer circles does set B have?
Compare the two sets to show 5 – 2 = 3.

The **difference** between the two sets is 3 because 3 of the circles in set A cannot be matched with set B.

Have students model several problems for finding the difference using sets from concrete manipulatives and pictorial representations.

Alternate Activity

○○ Have students look at 2 circles and think:

○○○○○ "How many more are needed to make 5?"

Students will think 2 + 3 = 5 and 5 – 2 = 3.

Have students model several problems demonstrating how many more of an item are needed. They can use concrete manipulatives and pictorial representations.

STUDENT PAGE
39
Subtraction

Model with a Number Line

Show students how to count backward on a number line.

EXAMPLE 5 – 2 = 3

1 Start at 5.

2 Count backward: –1 is 4, –2 is 3.

Have students practice several subtraction problems on the number line. Then use number lines that represent larger numbers.

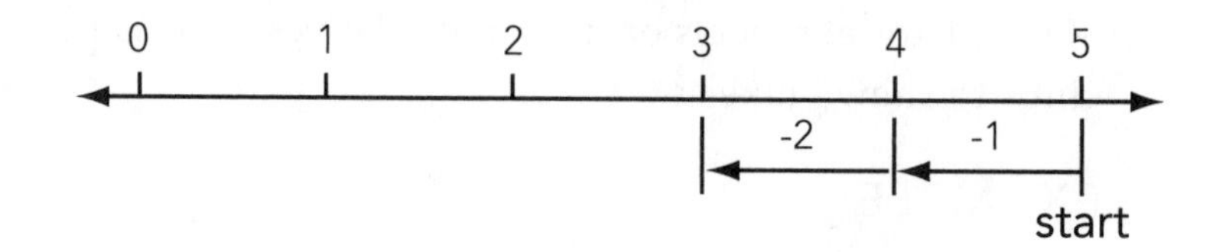

STUDENT PAGE
40
Subtraction

Create a Story

Ask students to make up their own stories from a number sentence written on the chalkboard. This helps students understand the operations based on concrete ideas they already understand.

EXAMPLES

A 9 – 4 = 5
I have 9 dollars and owe Sue 4 dollars. After paying Sue, how much money will I have left?

B 9 – ______ = 5
I have 9 dollars in my wallet. After I pay my dad what I owe him, I will have 5 dollars left. How much do I owe my dad?

C ______ – 4 = 5
I have 5 coins in my pocket after giving Anna 4. How many coins did I start with?

STUDENT PAGE
41
Subtraction

Counting Backward

Counting backward helps students understand subtraction. This kind of drill can be given orally or in written form.

EXAMPLES

Count backward from 50 by 5.
50, 45, 40, ______, ______, ______, ______, ______, ______, ______, 0

Count backward from 49 by 7.
49, 42, 35, ______, ______, ______, ______, 0

STUDENT PAGE
42
Subtraction

Relating Addition and Subtraction

Students can make several representations of two sets by using concrete manipulatives or drawing pictorial models. Once they have made the sets, have them write four related number sentences involving addition and subtraction based on the sets.

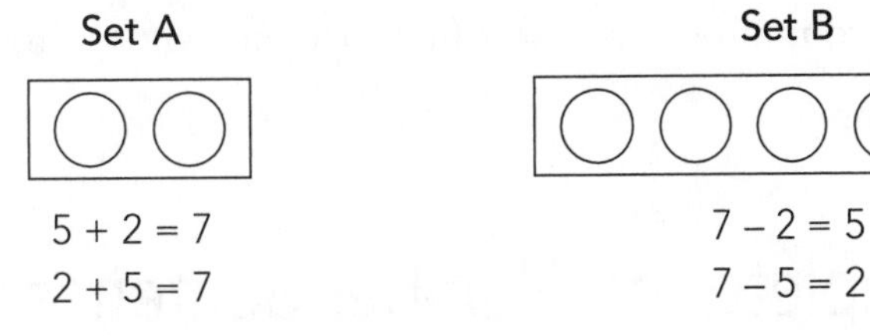

5 + 2 = 7	7 – 2 = 5
2 + 5 = 7	7 – 5 = 2

Alternate Activity

Ask students to write their own questions and complete number sentences based on them.

EXAMPLE What would you add to $18 to get $75?

$18 + ______ = $75

Alternate Activity

Write an addition problem on the board and have your students write and solve two related subtraction problems.

EXAMPLES

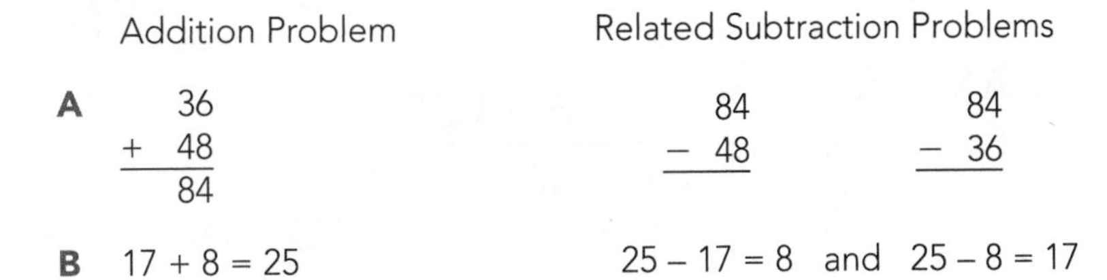

	Addition Problem	Related Subtraction Problems	
A	36 + 48 = 84	84 − 48	84 − 36
B	17 + 8 = 25	25 – 17 = 8 and 25 – 8 = 17	

STUDENT PAGE
42
Subtraction

Guess and Check

Make up problems that require the students to think about the relationship between two numbers.

EXAMPLES

Find two numbers whose sum is 13 and whose difference is 3. (Answer: 8 and 5)

Find two numbers whose difference is 8 and whose sum is 22. (Answer: 15 and 7)

Alternate Activity

Use problems similar to the examples that have three or four answers from which to select. If the students do not have to solve the problem, many will select the correct answer using mental computation.

EXAMPLE

Find two numbers whose sum is 18 and whose difference is 4.

A 8 and 5 **B** 9 and 7 **C** 7 and 11

STUDENT PAGE **47** Subtraction

Complete the Number Sentence

List four numbers on the board and have students mentally complete each number sentence using only those numbers. The degree of difficulty can be varied by changing the size of the numbers.

EXAMPLE [28] [9] [35] [13]

22 = ______ – ______ 7 = ______ – ______ 4 = ______ – ______

19 = ______ – ______ 15 = ______ – ______ 26 = ______ – ______

STUDENT PAGE **50** Subtraction

Scrambled Numbers

Scramble a series of numbers in a box and have your students list as many pairs as they can find whose difference is 25.

EXAMPLE

Find 25

35	10	28	
5	30	26	
40	15	3	1

STUDENT PAGE **53** Subtraction

Tell a Story

Have students formulate story problems from their personal real-life experiences.

EXAMPLE

The total enrollment of the high school is 525 students. 289 students bought tickets for the basketball game. How many students did not buy tickets?

STUDENT PAGE **54** Subtraction

Biggest Difference Wins

1 From a set of cards marked 0–9, each player or team takes a turn choosing one random card at a time. After each draw, the card is again mixed in for the next draw.

2 The object of the game is to build numbers that can be subtracted to result in the largest possible number. The level of difficulty can be adjusted by using different formats.

EXAMPLE FORMATS

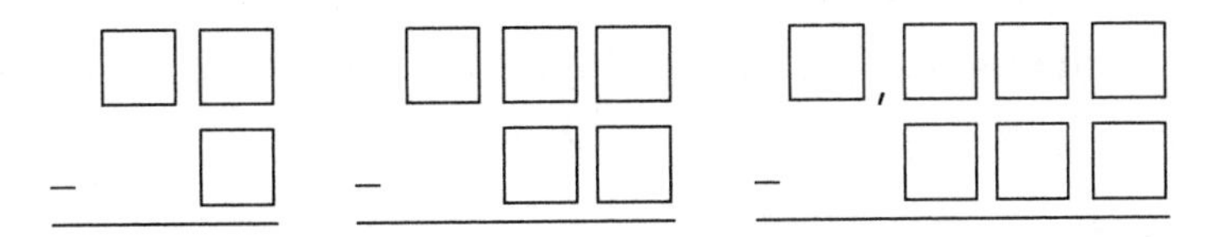

Alternate Activity

The player or team with the smallest number wins.

STUDENT PAGE
61
Subtraction

Find the Missing Digits

This kind of problem offers a good review of adding and subtracting whole numbers. Show students how to make new subtraction and addition problems to find the missing numbers.

5 □ □ − 8 5 4 7 5	1, 0 0 0 − □ □ □ 7 4 6	□ □ □ − 1 4 5 2 4 9
4 7 5 + 8 5 5 □ □	1, 0 0 0 − 7 4 6 □ □ □	2 4 9 + 1 4 5 □ □ □

STUDENT PAGE
62
Subtraction

Using Reference Points

Draw a diagram and let students decide which operations to perform to fill in the boxes with the correct number. Demonstrate how to do this before students do these independently or in small groups.

EXAMPLES

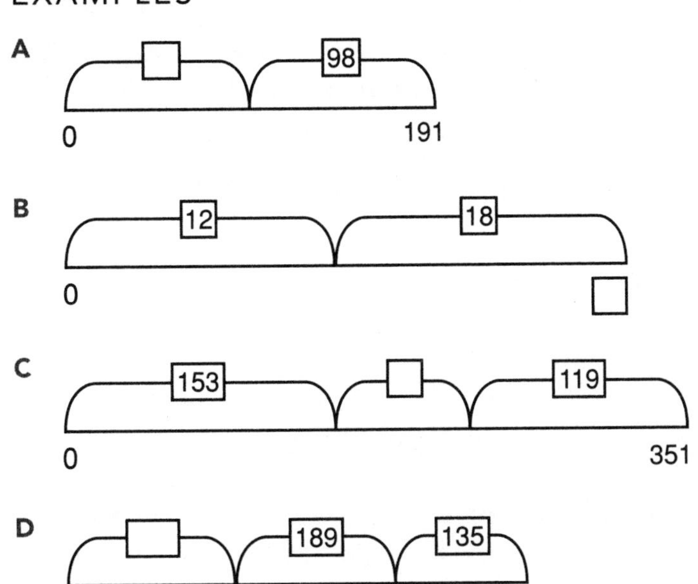

STUDENT PAGE
54
Subtraction Problem Solving

Draw a Picture

Show students that sketching reference points is a good problem-solving strategy.

EXAMPLES

A The flower shop has 95 roses. 36 are red roses. How many roses are not red?

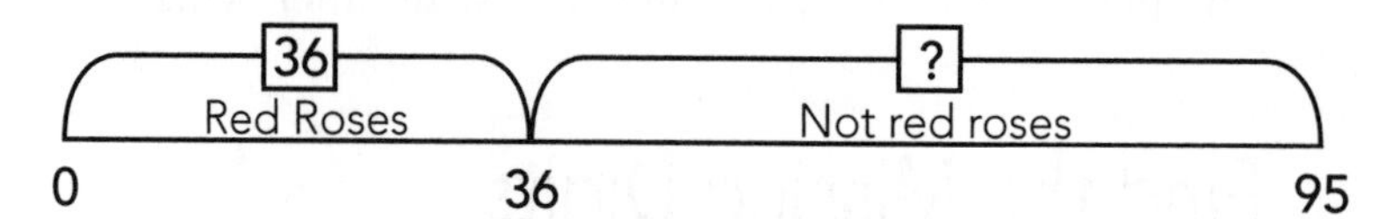

B A new radio costs $125. You have $75. How much more money do you need to buy the radio?

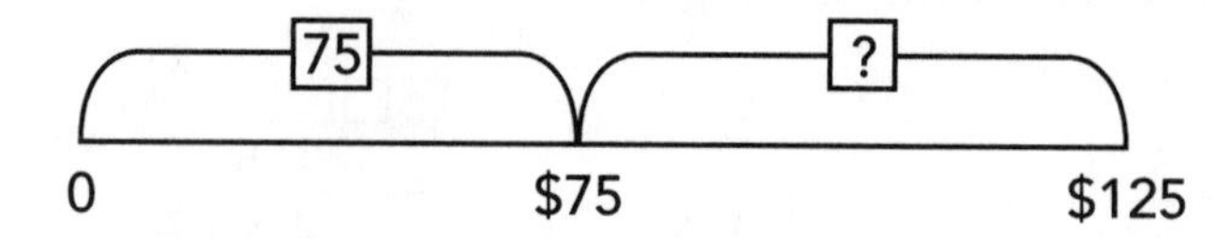

STUDENT PAGE
64
Subtraction Problem Solving

High-Interest Problem Solving

Let your students become active participants in creating and solving real-life problems.
Have students:

1. Fill in the blanks with reasonable numbers.
2. Write a question about number facts.
3. Complete a number sentence for each problem.
4. Write the answer in sentence form. This helps students clarify their thinking—"Does the answer make sense?"

Problem: There are *105* general admission seats and *200* reserved seats.
Question: *How many more reserved seats are there?*

200	−	105	=	95
Number	Operation Symbol	Number		Answer

Answer: *95 seats*

EXAMPLES

A Steve rode his bike ________ miles last week and ________ miles this week.

B ________ students attended the party. ________ wanted ice cream.

C Roger had ________ in his checking account. He wrote a check for ________.

D Ricardo had ________ baseball cards. Alicia had ________ cards.

STUDENT PAGE
67
Subtraction Problem Solving

Using Math Language

Orally drill your students using addition and subtraction math terms.

What number is 30 **more than** 25?
Combine 9 and 7.
15 **less than** 30 is what number?
The **difference** of 15 and 7 is what number?

STUDENT PAGE
65
Subtraction Word Problem

Word Sentences to Number Sentences

Students should learn to translate word sentences to number sentences to become familiar with the language of mathematics.

EXAMPLES

Word Sentences	Number Sentences
9 **plus** 7 **is equal to** 16	9 + 7 = 16
15 **minus** 6 **is equal to** 9	15 – 6 = 9
15 **is less than** 19	15 < 19
5 **plus** 6 **is equal to** 7 **plus** 4	5 + 6 = 7 + 4
9 **is greater than** 3	9 > 3

Alternate Activity

Students can translate number sentences to word sentences.

EXAMPLES

Number Sentences	Word Sentences
14 + 7 = 21	14 **plus** 7 **is equal to** 21
17 – 11 = 6	17 **minus** 11 **is equal to** 6
5 + 4 > 3	5 **plus** 4 **is greater than** 3

Page 4: Pretest

1. 107
2. 5,429
3. 1,669
4. 7,212
5. 121
6. 1,081
7. 135
8. 84
9. 26
10. 569
11. 379
12. 34
13. 108
14. 7,619
15. $275
16. $520
17. $2,605
18. 650 calories
19. 196 miles
20. 1,691 miles

Pretest Evaluation Chart

Skill Area	Pretest Problem #	Skill Section	Review Page
Addition	1, 2, 3, 4, 5, 6, 7	17–31	32
Subtraction	8, 9, 10, 11, 12, 13, 14	39–60	61
Addition Problem Solving	16, 18, 19	33–37	38
Subtraction Problem Solving	15, 17, 20	63–68	69
Life-Skills Math	All	70–73	74

Page 7: Hundreds

1. a) 8
   b) 9
2. a) 7
   b) 2
3. a) 4
   b) 0
   c) 5
4. a) 9
   b) 1
   c) 3
5. a) 3
   b) 0
   c) 0

Page 8: Thousands

1. a) 6
   b) 0
   c) 9
   d) 3
2. a) 4
   b) 3
   c) 6
   d) 1
   e) 2
3. a) 2
   b) 6
   c) 4
   d) 8
   e) 0
   f) 2

Page 9: Place-Value Readiness

Place-Value Chart

Chart: ten thousands, thousands, hundreds, tens, ones

1. thousands
2. tens
3. hundreds
4. ten thousands
5. ones
6. 5
7. 3
8. 8
9. 7
10. 2

Page 10: Number Sense

	BILLIONS			MILLIONS			THOUSANDS			UNITS		
	Hundred Billions	Ten Billions	Billions	Hundred Millions	Ten Millions	Millions	Hundred Thousands	Ten Thousands	Thousands	Hundreds	Tens	Ones
1.										6	8	9
2.									6	0	9	3
3.							1	4	9	5	0	8
4.								2	3	1	5	4
5.						5	9	0	4	1	6	5
6.										1	9	6
7.					6	2	9	4	9	1	5	0
8.									3	9	9	2
9.				6	4	9	1	3	2	0	0	1
10.	3	2	6	1	9	4	6	0	0	1	0	9

Page 11: Commas

A. 346
B. 12,346
C. 21,612,346
D. 7,421,612,346

1. 7,349
2. 21,496
3. 14,948
4. 964,821
5. 1,436,849
6. 29,743,268
7. 1,413,962
8. 217,924
9. 1,439
10. 341
11. 11,682,371
12. 46,218

Page 12: Words to Numbers

1. 509
2. 703
3. 2,116
4. 5,008
5. 6,843
6. 74,904
7. 7,004
8. 317,052,000
9. 29,704,500,396
10. 9,015
11. 20,045
12. 30,436,165

Page 13: Expanded Forms

1. (7 × 10) + (9 × 1)
2. (1 × 100) + (9 × 10) + (6 × 1)
   100 + 90 + 6
3. (6 × 1,000) + (0 × 100) + (9 × 10) + (3 × 1)
   6,000 + 0 + 90 + 3
4. (4 × 1,000) + (9 × 100) + (8 × 10) + (0 × 1)
   4,000 + 900 + 80 + 0
5. (5 × 1,000) + (1 × 100) + (4 × 10) + (6 × 1)
   5,000 + 100 + 40 + 6

Page 14: Ordering Numbers

1. 83 115 380 651 4,022
2. 1,094 802 290 136 75
3. 158
4. 237
5. 249
6. 7,420
7. 8,532
8. 9,610
9. 75,430
10. 95,431

Page 15: Comparing Numbers

1. 48
2. 513
3. 9,048
4. 55,246
5. <
6. >
7. >
8. =
9. >
10. >
11. >
12. >

Page 16: Place Value Review

1. hundreds
2. hundred thousand
3. 6
4. 2,394,738
5. 159,642
6. (4 × 1,000) + (8 × 100) + (2 × 10) + (6 × 1)
7. 249
8. 96,542
9. <
10. <

Page 17: Meaning of Addition

A. 2 + 4 = 6
B. 6
1. 4 + 5 = 9
2. 5 + 7 = 12
3. 3 + 5 = 8
4. 5 + 8 = 13

Page 18: Addition Facts

+	+0	+1	+2	+3	+4	+5	+6	+7	+8	+9
1	1	2	3	4	5	6	7	8	9	10
2	2	3	4	5	6	7	8	9	10	11
3	3	4	5	6	7	8	9	10	11	12
4	4	5	6	7	8	9	10	11	12	13
5	5	6	7	8	9	10	11	12	13	14
6	6	7	8	9	10	11	12	13	14	15
7	7	8	9	10	11	12	13	14	15	16
8	8	9	10	11	12	13	14	15	16	17
9	9	10	11	12	13	14	15	16	17	18

Page 19: Practice Helps

1. 14
2. 11
3. 15
4. 11
5. 13
6. 16
7. 15
8. 13
9. 12
10. 16
11. 13
12. 17
13. 13
14. 6
15. 14
16. 15
17. 13
18. 10
19. 11
20. 16
21. 17
22. 16
23. 10
24. 15
25. 18
26. 17
27. 13
28. 16
29. 12
30. 17
31. 14
32. 14
33. 12
34. 12
35. 15
36. 7
37. 16
38. 13
39. 11
40. 15
41. 11
42. 16
43. 14
44. 12
45. 12
46. 15
47. 18
48. 13
49. 11
50. 15
51. 11
52. 14
53. 12
54. 17
55. 9
56. 8
57. 10
58. 10
59. 11
60. 12
61. 12
62. 15
63. 16
64. 14
65. 13
66. 11
67. 15
68. 13
69. 13
70. 12
71. 14
72. 14
73. 11
74. 4
75. 9
76. 10
77. 12
78. 10
79. 16
80. 14

Page 20: Counting Patterns

1. 5, 10, 15, 20, 25, 30, 35, 40, 45, 50, 55, 60
2. 10, 20, 30, 40, 50, 60, 70, 80, 90, 100, 110, 120
3. 3, 5, 7, 9, 11, 13, 15, 17, 19, 21, 23, 25
4. 13, 19, 25, 31, 37, 43, 49, 55, 61, 67, 73, 79
5. 5, 9, 13, 17, 21, 25, 29, 33, 37, 41, 45, 49
6. 0, 3, 6, 9, 12, 15, 18, 21, 24, 27, 30, 33
7. 3, 10, 17, 24, 31, 38, 45, 52, 59
8. 4 + 9, 13 + 9, 22, 31, 40, 49, 58, 67, 76
9. 9 + 9, 18 + 9, 27, 36, 45, 54, 63, 72, 81
10. 7, 15, 23, 31, 39, 47, 55, 63, 71
11. 11, 17, 23, 29, 35, 41, 47, 53, 59
12. 9, 16, 23, 30, 37, 44, 51, 58, 65

Page 21: Timed Addition Drill

1. 14
2. 18
3. 20
4. 19
5. 9
6. 10
7. 10
8. 14
9. 10
10. 10
11. 12
12. 15
13. 11
14. 14
15. 16
16. 16
17. 13
18. 17
19. 7
20. 11
21. 9
22. 6
23. 9
24. 9
25. 17
26. 8
27. 15
28. 14
29. 19
30. 11
31. 11
32. 13
33. 11
34. 8
35. 12
36. 14
37. 16
38. 12
39. 18
40. 8
41. 7
42. 6
43. 15
44. 13
45. 12
46. 12
47. 17
48. 12
49. 16
50. 18
51. 5
52. 9
53. 15
54. 7

Page 22: Column Addition

1. 21
2. 20 [14] + 6
3. 17 [9] + [8]
4. 14 [9] + [5]
5. 20
6. 23
7. 24
8. 26
9. 17
10. 19
11. 21
12. 35 [13]

Page 23: Two-Digit Addition

A. 95
B. 83
C. 78
1. 39
2. 53
3. 87
4. 99
5. 99
6. 48
7. 48
8. 60
9. 98
10. 89
11. 59
12. 77
13. 67
14. 77
15. 88
16. 78

Page 24: Three-Digit Addition

1. 979 **2.** 848 **3.** 474 **4.** 505
5. 498 **6.** 698 **7.** 531 **8.** 587
9. 947 **10.** 899 **11.** 373 **12.** 849
13. 935 **14.** 249 **15.** 586 **16.** 575

Page 25: Find the Missing Number

1. 2 **2.** 3 **3.** 0 **4.** 5 **5.** 3 **6.** 2 **7.** 5 **8.** 0 **9.** 3 **10.** 5
11. 4 **12.** 4 **13.** 5 **14.** 4 **15.** 3 **16.** 6 **17.** 6 **18.** 2 **19.** 2 **20.** 4
21. 3 **22.** 5 **23.** 0 **24.** 1 **25.** 8 **26.** 7 **27.** 5 **28.** 4 **29.** 4 **30.** 0
31. 3 **32.** 3 **33.** 4 **34.** 3 **35.** 6 **36.** 7 **37.** 6 **38.** 3 **39.** 2 **40.** 3
41. 3 **42.** 2 **43.** 5 **44.** 8 **45.** 0 **46.** 2 **47.** 1 **48.** 2 **49.** 5 **50.** 1
51. 4 **52.** 1 **53.** 6 **54.** 2 **55.** 3 **56.** 3 **57.** 0 **58.** 2 **59.** 0 **60.** 5
61. 1 **62.** 2 **63.** 3 **64.** 1 **65.** 3 **66.** 4 **67.** 0 **68.** 0 **69.** 3 **70.** 3
71. 8 **72.** 1 **73.** 4 **74.** 3 **75.** 0 **76.** 5 **77.** 6 **78.** 6 **79.** 1 **80.** 3

Page 26: Regroup the Ones

1. 53 **2.** 95 **3.** 92 **4.** 131
5. 123 **6.** 153 **7.** 120 **8.** 117
9. 134 **10.** 161 **11.** 110 **12.** 111
13. 185 **14.** 160 **15.** 136 **16.** 131

Page 27: Regroup and Think Zero

A. 62 **B.** 105 **C.** 80
1. 42 **2.** 100 **3.** 31
4. 20 **5.** 51 **6.** 88 **7.** 50 **8.** 84 **9.** 61
10. 72 **11.** 31 **12.** 74 **13.** 50 **14.** 74 **15.** 60
16. 75 **17.** 41 **18.** 83 **19.** 101 **20.** 23

Page 28: Adding More Numbers

1. 1,478 **2.** 1,311
3. 1,773 **4.** 1,290
5. 1,412 **6.** 1,770

Page 29: Addition Practice

1. 68 **2.** 160 **3.** 169 **4.** 190
5. 879 **6.** 1,123 **7.** 1,746 **8.** 15,320

Page 30: Using a Grid

1.

9,	3	0	4
		9	2
	6	4	5

2.

		3	2
	1	9	5
6,	4	5	2

3.

8,	8	9	4
		6	5
3,	2	0	5

4.

		3	9
2,	5	0	3
			4

5.

		9	5
			6
	1	3	9
6,	4	8	0

6.

	8,	9	5	6
8	8,	3	0	9
				6

7.

2	3,	9	1	4
			5	6
				9
			1	3

8.

		3	9	2
				8
9	4,	1	3	8
			1	4
				4

Page 31 Lining Up Numbers

1. 13,514
2. 7,879
3. 1,110
4. 8,320
5. 7,111
6. 6,290
7. 1,952
8. 9,889

Page 32: Addition Review

1. a) ④,932 b) 60⑨,056 c) 5④,962
2. a) ③49 b) 9,⑥14 c) 32,①00
3. a) 8④ b) 14⑥ c) 2,29⑥
4. a) ⑨3 b) 6,8④0 c) 61,2③2
5. 75 431 705 9,103 47,147
6. 1,932 = (1 × 1,000) + (9 × 100) + (3 × 10) + (2 × 1) = 1,000 + 900 + 30 + 2
7. 4,379,321
8. 197
9. 8,023
10. 3,212

Page 33: Addition Problems

1. 3 + 2 = 5
2. 5 + 3 = 8
3. 6 + 4 = 10
4. answers will vary: 4 + 4 = 8
5. 4 + 5 = 9
6. 13 + 8 = 21

Page 34: Problem-Solving Strategies

1. $9 + $5 = $14
2. $17 + $15 = $32
3. 19 + 8 = 27
4. 13 + 7 = 20
5. 8 + 9 = 17
6. 16 + 9 = 25
7. 6 + 3 = 9
8. $4 + $3 = $7

Page 35: Addition Word Problems

1. 128 + 75 = 203
2. $19,954 + $2,129 = $22,083
3. 6,495 + 5,092 = 11,587
4. 486 + 294 = 780
5. 24,182 + 6,759 = 30,941
6. 48,560 + 25,709 = 74,269

Page 36: Think About the Facts

Answers may be similar to these.

1. How many CDs did Rusty buy?
   8 + 3 = 11
2. How much money does Nicole have?
   $73 + $52 = $125
3. How far did Terry drive?
   120 + 140 = 260
4. How many cups did they buy altogether?
   144 + 72 = 216
5. How many pairs of earrings does Shannon have?
   57 + 38 = 95
6. How much did they earn altogether?
   $855 + $875 = $1,730

Page 37: Using Symbols

1. a) $37 + 52 = 89$
   b) $52 > 37$
   c) $37 < 52$
   d) $37 \neq 52$
2. a) $75 + 80 = 155$
   b) $50 + 100 = 150$
   c) $75 > 50$
   d) $80 < 100$
   e) $155 \neq 150$

Page 38: Word Problem Review

1. 212 + 59 = 271
2. \$18 + \$32 = \$50
3. 87 + 15 = 102
4. \$12 + \$18 = \$30
5. 117 + 89 = 206
6. 314 + 189 = 503
7. 50 + 75 = 125
8. \$4,569 + \$557 = \$5,126
9. \$550 + \$749 = \$1,299
10. \$139 + \$99 = \$238

Page 39: Meaning of Subtraction

1. a) 5 b) 8 − 3 = 5
2. a) 3 b) 5 − 2 = 3
3. 6 − 3 = 3

Page 40: Practice Helps

1. 1	21. 0	41. 7	61. 4
2. 2	22. 8	42. 7	62. 9
3. 5	23. 1	43. 8	63. 8
4. 4	24. 0	44. 4	64. 6
5. 0	25. 3	45. 5	65. 9
6. 2	26. 2	46. 5	66. 5
7. 1	27. 0	47. 8	67. 10
8. 2	28. 2	48. 9	68. 7
9. 2	29. 0	49. 8	69. 9
10. 5	30. 3	50. 4	70. 10
11. 8	31. 1	51. 6	71. 11
12. 1	32. 6	52. 4	72. 5
13. 5	33. 6	53. 2	73. 3
14. 3	34. 1	54. 8	74. 7
15. 5	35. 3	55. 7	75. 9
16. 4	36. 2	56. 7	76. 3
17. 3	37. 6	57. 6	77. 10
18. 4	38. 7	58. 3	78. 10
19. 3	39. 4	59. 7	79. 5
20. 2	40. 5	60. 12	80. 6

Page 41: Timed Subtraction Drill

1. 3	15. 6	29. 1	43. 9
2. 3	16. 1	30. 3	44. 5
3. 7	17. 0	31. 3	45. 5
4. 0	18. 2	32. 10	46. 8
5. 4	19. 3	33. 5	47. 4
6. 1	20. 2	34. 6	48. 2
7. 0	21. 3	35. 8	49. 8
8. 7	22. 4	36. 0	50. 10
9. 4	23. 4	37. 7	51. 1
10. 1	24. 2	38. 11	52. 6
11. 4	25. 5	39. 7	53. 8
12. 0	26. 3	40. 4	54. 6
13. 8	27. 4	41. 11	
14. 1	28. 5	42. 0	

Page 42: Relating Addition and Subtraction

1. 10 − 4 = 6
2. 14 − 5 = 9
3. 7 − 4 = 3
4. 6 − 1 = 5
5. 7 − 7 = 0
6. 13 − 8 = 5
7. 4 + 7 = 11, so 11 − 4 = 7
8. 7 + 3 = 10, so 10 − 3 = 7
9. 3 + 9 = 12, so 12 − 9 = 3
10. 9 + 0 = 9, so 9 − 0 = 9
11. 2
12. 9
13. 15
14. 6
15. 8
16. 24
17. 14
18. 97
19. 29
20. 8

Page 43: Subtracting Ones and Tens

1. a) 27 b) add
2. 32
3. 22
4. 30
5. 51
6. 12
7. 22

Page 44: Think Zero

A. 33	**5.** 72	**12.** 31
B. 22	**6.** 31	**13.** 56
C. 72	**7.** 20	**14.** 20
1. 42	**8.** 55	**15.** 32
2. 62	**9.** 24	**16.** 41
3. 41	**10.** 81	
4. 34	**11.** 55	

Page 45: Regrouping to Subtract

1. 13	**3.** 12	**5.** 18
2. 16	**4.** 15	**6.** 10

Page 46: Subtraction Readiness

1. 7 tens 4 ones 74
2. 5 tens 9 ones 59
3. 4 tens 8 ones 48
4. 4 tens 9 ones 49
5. 6 tens 9 ones 69
6. 5 tens 6 ones 56

Page 47: Regrouping Tens to Ones

Step 4 29

1. 19	**4.** 88	**7.** 47	**10.** 29
2. 18	**5.** 57	**8.** 58	**11.** 45
3. 79	**6.** 58	**9.** 27	**12.** 68

Page 48: Regroup and Subtract

1. 34	**4.** 79	**7.** 44	**10.** 39
2. 19	**5.** 36	**8.** 6	**11.** 27
3. 9	**6.** 29	**9.** 19	**12.** 28

Page 49: Deciding to Regroup

1. Yes; 8 is larger than 7
2. Yes; 7 is larger than 6
3. No; 4 is smaller than 9
4. No; 1 is smaller than 9
5. Yes; 7 is larger than 5
6. No; 1 is smaller than 9

Page 50: Regroup Only When Necessary

1. 14	**4.** 44	**7.** 39	**10.** 40
2. 44	**5.** 24	**8.** 21	**11.** 59
3. 36	**6.** 5	**9.** 15	**12.** 16

Page 51: Subtract and Check

1. 413	**3.** 227	**5.** 169	**7.** 209
2. 229	**4.** 129	**6.** 439	**8.** 219

Page 52: Regrouping Tens and Hundreds

1. 652	**4.** 582	**7.** 374
2. 291	**5.** 691	**8.** 195
3. 173	**6.** 245	**9.** 702

Page 53: Regrouping More Than Once

1. 383	**4.** 189	**7.** 93	**10.** 279
2. 387	**5.** 169	**8.** 78	**11.** 177
3. 177	**6.** 469	**9.** 259	**12.** 179

Page 54: Regrouping from Zero

A. 335
1. 131
2. 37
3. 184
4. 370
5. 222
6. 416
7. 243
8. 462
9. 204
10. 851
11. 40
12. 501

Page 55: Larger Numbers

A. 4,325
B. 7,736
1. 3,065
2. 1,754
3. 8,538
4. 5,493
5. 5,373
6. 1,154
7. 4,938
8. 4,921
9. 6,409
10. 3,284
11. 2,509
12. 4,948

Page 56: Subtraction Practice

1. 21
2. 16
3. 422
4. 469
5. 111
6. 59
7. 1,262
8. 6,717

Page 57: Using a Grid

A.

	9	8	4
−		9	3

1.

	4	5	1	5	4
−			3	4	2

2.

	6	4	5
−		1	5

3.

	1	9	9	5
−		8	8	4

4.

	6	8
−		7

5.

	2	1	6	7	3
−		1	8	4	2

6.

	8	9	1	5	0	5	6
−		1	2	5	6	0	4

7.

	2	0	1	9
−		3	4	0

8.

	2	3	5
−	1	3	3

Page 58: Line Up to Subtract

1. 243
2. 462
3. 355
4. 415
5. 4,548
6. 2,350
7. 4,184
8. 3,988
9. 2,660
10. 2,753
11. 51,510
12. 45,044

Page 59: Take Away

A. 61
B. 1,029
1. 66
2. 463
3. 127
4. 289
5. 1,917
6. 647

Page 60: Find the Difference

A. 370
B. 609
1. 225
2. 539
3. 316
4. 177
5. 479
6. 483

Page 61: Subtraction Review

1. If 6 + 7 = 13, then 13 − 7 = 6
2. 83
3. 163
4. Yes; 9 is larger than 8.
5. 457
6. 379
7. 131
8. 566
9. 368
10. 1,829

Page 62: Putting It All Together

1. <
2. =
3. <
4. <
5. >
6. =
7. =
8. <
9. >
10. =
11. >
12. =
13. =
14. <

Page 63: Subtraction Problems

1. 5 – 2 = 3
2. 5 – 2 = 3
3. 6 – 4 = 2
4. 7 – 3 = 4
5. a) 2 b) 2 c) 3 – 1 = 2
6. a) 6 b) 6 c) 10 – 4 = 6

Page 64: Problem-Solving Strategies

1. 7 – 3 = 4
2. 14 – 8 = 6
3. 15 – 7 = 8
4. 25 – 14 = 11
5. 95 – 36 = 59
6. $38 – $10 = $28
7. $125 – $75 = $50
8. $25 – $12 = $13

Page 65: Subtraction Word Problems

1. 98 – 53 = 45
2. 84 – 36 = 48
3. $125 – $95 = $30
4. 35 – 16 = 19
5. 78 – 15 = 63
6. $194 – $75 = $119
7. $545 – $470 = $75
8. $215 – $175 = $40
9. $75 – $15 = $60
10. $75 – $32 = $43

Page 66: Think About the Facts

Answers may be similar to these.

1. How many students are there in all?
   39 + 16 = 55
2. What is the total of her car and rent payments?
   $260 + $195 = $455
3. How much money does she have after paying bills?
   $1,537 – $895 = $642
4. How many miles did he drive in all?
   198 + 239 = 437
5. How many empty seats were there?
   497 – 239 = 258
6. How many cards do they have altogether?
   125 + 65 = 190

Page 67: Using Symbols

1. a) 21 – 6 = 15
   b) 15 + 6 = 21
   c) 21 > 15
   d) 15 < 21
   e) $15 \neq 21$
2. a) 16 – 7 = 9
   b) 9 + 7 = 16
   c) 16 > 9
   d) 9 < 16
   e) $16 \neq 9$

Page 68: Practice with Symbols

1. a) 3 + 7 = 10
   b) 10 – 7 = 3
   c) 10 > 3
   d) 3 < 10
   e) $10 \neq 3$
2. a) 30 + 5 = 35
   b) 35 – 5 = 30
   c) 35 > 30
   d) 30 < 35
   e) $35 \neq 30$

Page 69: Word Problem Review

1. 42 + 57 = 99
2. $20 − $14 = $6
3. 28 − 17 = 11
4. $9 + $15 = $24
5. $125 + $250 = $375
6. 250 + 138 = 388
7. 704 + 435 = 1,139
8. 235 − 68 = 167
9. 245 − 190 = 55
10. 136 − 25 = 111

Page 70: Picture Problems

1. $57
2. $133
3. 37 miles
4. 94 miles
5. $144
6. a) 8 hours
   b) 3 hours
   c) 4 hours
   d) 6 hours
   e) 5 hours
7. 26 hours

Page 71: Reading a Map

1. a) 49 miles
   b) 52 miles
   c) 66 miles
   d) 74 miles
   e) 89 miles
   f) 64 miles
   g) 49 miles
2. 202 miles
3. 140 miles
4. 153 miles
5. a) Detroit to Kalamazoo
   b) 13 miles
6. $3,491

Page 72: Using Checks and Calendars

1. Answers should reflect a current date.
2. $196
3. One hundred fifty-eight and 00/100
4. $38
5. Sunday
6. June 2, June 16, and June 30
7. June 26
8. 14 days

Page73: Comparing Prices

1. a) $1,273
   b) $1,268
2. $1,265
3. $764
4. $501
5. a) Answers will vary.
   b) Answers will vary.

Page 74: Life-Skills Math Review

1. $8
2. 343 tickets
3. 33 miles
4. $162
5. $15,458

Page 75: Cumulative Review

1. 684
2. 855
3. 28
4. $8
5. 216

Page 76: Posttest

1. 101
2. 58
3. 27,174
4. $1,576
5. 2,482
6. 6,611
7. 3,932
8. 6,709
9. 39,251
10. 981
11. 226 years old
12. 396
13. 6,633
14. 748
15. 5,903
16. 1,359 miles
17. $320
18. 1,403
19. $27,625
20. $88

Posttest Evaluation Chart

Skill Area	Pretest Problem #	Skill Section	Review Page
Addition	1, 3, 5, 7, 10, 13	17–31	32
Subtraction	6, 8, 12, 14, 15, 18	39–60	61
Addition Word Problems	4, 9, 17, 19	33–37	38
Subtraction Word Problems	2, 11, 16, 20	63–68	69
Life-Skills Math	All	70–73	74

WHOLE NUMBERS
Multiplication & Division

The approach for teaching multiplication and division is similar to that used for addition and subtraction. Teachers should develop an understanding of the concepts by proceeding from the concrete to the abstract and symbolic.

Student Glossary

Acquainting students with definitions of key math terms and life-skills concepts will enhance their mastery of the materials. Below are words defined in the student text. A glossary is provided at the end of the student text and on page 243 of the *Teacher's Resource Guide and Answer Key.*

clockwise
dividend
division
divisor
estimate
multiple
multiplication
number relation symbol
operation symbol
place value
quantity
quotient
regroup (carry)
regular price
remainder
round
sale price
shift
standard time
subscription
times
unit price

STUDENT PAGE
7
Multiplication

Multiplication Using Manipulatives

Use concrete manipulatives to demonstrate sets by replacing Xs with blocks, straws, clothespins on cards, etc.

EXAMPLE

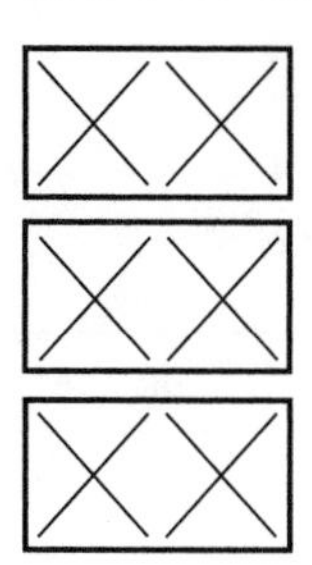

Have the students determine:

How many groups of 2? 3

How many Xs in each group? 2

How many Xs in all? 6

Summarize: $3 \times 2 = 6$

Alternate Activity

You can make up stories about the manipulatives. This will make the multiplication fact concrete to the students.

EXAMPLE

We need 2 pencils for each student. How many pencils will we need for 3 students?

When a student gets the correct answer, ask how the student arrived at the answer.

Alternate Activity

Have students make up their own sets and write multiplication facts based on the sets.

EXAMPLE

There are two shelves in a medicine cabinet. Each shelf holds four jars. How many jars will the shelves hold all together?

2	×	4	=	8
groups		how many in each group?		total

This can be read as 2 sets of 4 equals 8.

Encourage students to tell or write their own stories based on the sets.

STUDENT PAGE
8
Multiplication

Multiplication Using a Number Line

The number line is another model that may be used with students. You can provide many examples modeling multiplication on the number line.

EXAMPLE

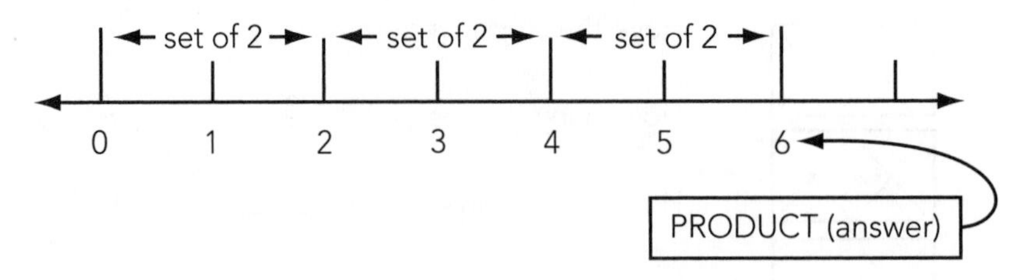

Summarize: 3 sets of 2 are 6, which shows that 3 x 2 = 6.

STUDENT PAGE
9
Multiplication

Sketching Groups

Have students sketch three Xs in each group.

Have the students determine:

XXX

How many groups? 6

How many Xs in each group? 3

How many Xs in all? 18

6	×	3	=	18
groups		how many in each group?		total

Summarize: **6 × 3 = 18.**

Have students take turns sketching Xs in different numbers of groups and asking each other questions. Point out that they must sketch the same number of Xs in each group or it will not be a multiplication problem.

STUDENT PAGE
10
Multiplication

Multiplication Using Arrays

Arrays are arrangements of objects or symbols in orderly rows and columns. They can be very convincing models for illustrating the commutative property of multiplication. The **commutative property** means that the order of the two factors being multiplied does not affect the product.

Provide students with graph paper and let them make up arrays for basic muliplication facts.

EXAMPLE $6 \times 3 = 3 \times 6$

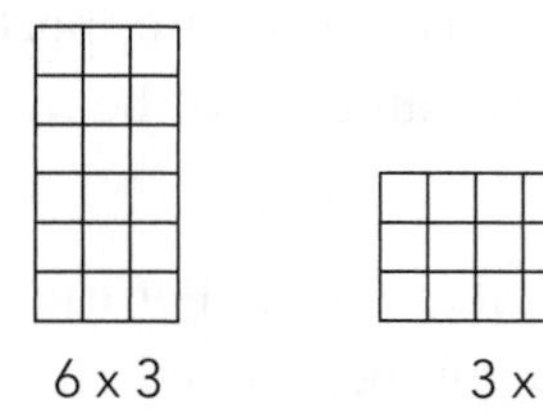

6 x 3 3 x 6

Alternate Activity

Geoboards can be used to demonstrate the commutative property of multiplication. Demonstrate an example for the students, then have students make their own designs to show their understanding.

EXAMPLE

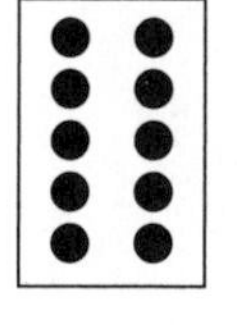

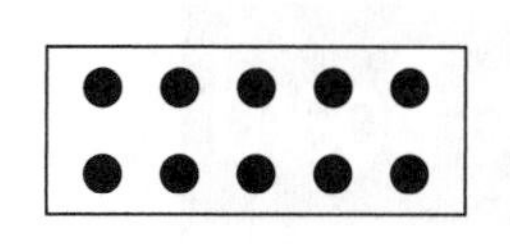

5 x 2 2 x 5

Use geoboards and rubber bands.

Use concrete manipulatives; e.g. blocks, dried beans, etc.

1 Make a picture of a multiplication fact using arrays. Do two arrays to demonstrate the commutative property.

2 Show that the order of **factors** makes no difference to the **product**.

STUDENT PAGE
11
Multiplication

Memorizing the Facts

Knowledge of the basic multiplication facts is essential. Students must commit all facts to memory. For those who do not know all their facts, develop drill games using flash cards, number cubes, races, etc.

EXAMPLE

Have students work through flash cards answering all the problems. When a student misses a fact, put that card aside. When all of the flash cards have been covered, go back over only those cards the student has missed. Repeat the process until all facts have been mastered.

STUDENT PAGE
12
Multiplication

Multiplication Bingo

On a blank 3 × 6 bingo board, have students scatter the numbers 1, 2, 3, 4, 5, 6, 8, 9, 10, 12, 15, 16, 18, 20, 24, 25, 30, and 36. It does not matter in which square the numbers are placed. This is a good opportunity to discuss **prime numbers** and why 13, 17, 19, etc. should not be included on their bingo board.

Each player takes turns rolling two number cubes. For each throw, the player calls out the product of the two numbers rolled and covers the answer on the board.

EXAMPLE

A player rolls a 3 and a 4. The player states that 3 x 4 = 12. The students should check their bingo board and cover 12 on the board.

Play continues until one player covers all the numbers in any one of the three vertical columns.

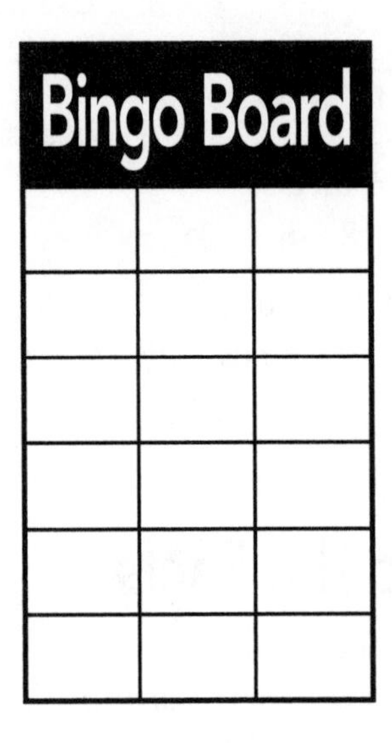

STUDENT PAGE
13
Multiplication

Multiplication Bingo

Students who have memorized multiplication facts through the number 3 have only 21 facts left to memorize. This game provides a good way to practice the 21 facts.

Have students sketch two 3 × 3 bingo boards on their paper and scatter 18 **products** (9 per board) from this list of 21 problems:

4 × 4	5 × 5	6 × 6	7 × 7	8 × 8	9 × 9
4 × 5	5 × 6	6 × 7	7 × 8	8 × 9	
4 × 6	5 × 7	6 × 8	7 × 9		
4 × 7	5 × 8	6 × 9			
4 × 8	5 × 9				
4 × 9					

It does not matter in what square the products are placed.

Scramble the order of the multiplication facts. Using flash cards or writing the facts on the board one at a time, have students cover the product on their bingo boards. Play continues until a player gets bingo by covering any row, column, or diagonal. Multiple bingo boards can be played at once.

Alternate Activity

Students sketch a 6 x 6 bingo board on their paper. The leader reads the products of the 36 facts listed below for students to scatter on their boards. The numbers 12, 18, and 36 must be used twice. Next, as the leader reads each multiplication fact, students cover or cross out each product on their card. Play continues until a player gets bingo by covering any row, column, or diagonal.

Students appreciate knowing there are only 10 facts to learn if they have mastered all facts through the number 5.

2 × 2 = 4	3 × 3 = 9	4 × 4 = 16	5 × 5 = 25
2 × 3 = 6	3 × 4 = 12	4 × 5 = 20	5 × 6 = 30
2 × 4 = 8	3 × 5 = 15	4 × 6 = 24	5 × 7 = 35
2 × 5 = 10	3 × 6 = 18	4 × 7 = 28	5 × 8 = 40
2 × 6 = 12	3 × 7 = 21	4 × 8 = 32	5 × 9 = 45
2 × 7 = 14	3 × 8 = 24	4 × 9 = 36	
2 × 8 = 16	3 × 9 = 27		
2 × 9 = 18			

6 × 6 = 36	7 × 7 = 49	8 × 8 = 64	9 × 9 = 81
6 × 7 = 42	7 × 8 = 56	8 × 9 = 72	
6 × 8 = 48	7 × 9 = 63		
6 × 9 = 54			

STUDENT PAGE
14
Multiplication

Around the World

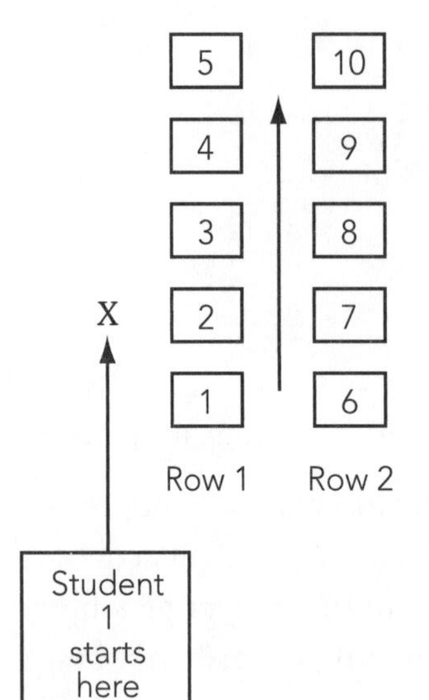

1. Student 1 stands next to the desk of Student 2.
2. The teacher reads a multiplication fact.
3. The student who first answers correctly moves on to stand beside the next desk in the row, and the student who answered incorrectly sits down. A student may keep moving up and down the rows as long as he or she answers correctly.
4. The student who answers correctly the most times is the winner.

STUDENT PAGE
14
Multiplication

Flash Cards

Place a row of flash cards along the top of the chalkboard. Time students to see how fast they can answer the row of multiplication facts. The student with the fastest time wins.

STUDENT PAGE
14
Multiplication

Review the Facts

X	3	7	9	
5	15	35		
6		42		
8				

To review basic multiplication facts, use a grid. You may add more numbers along the top and side.

STUDENT PAGE
19
Multiplication

Rounding Numbers for Super-Fast Answers

$$\begin{array}{r} 56 \\ \times\ 47 \\ \hline \end{array} \qquad \begin{array}{r} 60 \\ \times\ 50 \\ \hline 3{,}000 \end{array}$$

Teach students to round numbers to the nearest 10 and estimate the answer.

Have students compare their estimates with the exact answer.

- Were the estimated answers close?
- How close?

STUDENT PAGE
19
Multiplication

Estimate the Answer

Tell students that the ability to estimate answers will help them determine if an answer is accurate. First, teach students to round to the nearest 10, 100, and 1,000. On a worksheet or on the board, give students problems and estimated answer choices. Have them circle the best estimate for each problem.

Problem	Answer Choices		
38 × 59 =	240	2,400	24,000
734 × 9 =	634	6,340	63,400
884 × 59 =	540	5,400	54,000

Next, have students write their own problems and quiz each other.

STUDENT PAGE
20
Multiplication

Fill In the Missing Digits

This kind of problem offers a good review for multiplying whole numbers.

EXAMPLE

$$\begin{array}{r} 3\ 8\ 9 \\ \times\ \ 7\ \square \\ \hline \square\ \square\ 1\ 2 \\ \square\ \square\ \square\ 3\ 0 \\ \hline 3\ \square\ \square\ \square\ \square \end{array}$$

With this type of problem, teach students to use trial and error or guess and check methods.

SOLUTION

$$\begin{array}{r} 3\ 8\ 9 \\ \times\ \ 7\ \boxed{8} \\ \hline \boxed{3}\ \boxed{1}\ 1\ 2 \\ \boxed{2}\ \boxed{7}\ \boxed{2}\ 3\ 0 \\ \hline 3\ \boxed{0}\ \boxed{3}\ \boxed{4}\ \boxed{2} \end{array}$$

Guess and check.
Think about what number times 9 has an answer ending with 2.
Does 8 work?
9 × 8 = 72

STUDENT PAGE **25** Multiplication Problem Solving

Real-Life Story Problems

Let your students become active participants in creating and solving real-life problems.

Have students:

1. Fill in the blanks with reasonable numbers.
2. Write a question about the facts.
3. Complete a number sentence for each problem.
4. Write the answer in sentence form. This helps students clarify their thinking—"Does the answer make sense?"

EXAMPLE

Mario traveled __55__ miles per hour for __3__ hours.

Question: __How many miles did Mario travel?__

55	×	3	=	165
Number	Operation	Number		Answer

Answer: 165 miles

STUDENT PAGE **25** Division

Use Concrete Manipulatives

Use concrete manipulatives by replacing Xs with blocks, straws, clothespins, etc.

EXAMPLE

Ask:

How many Xs altogether?

How many Xs in each group?

How many groups?

Summarize: 15 ÷ 3 = 5

Alternate Activity

Make up stories to help students grasp division concepts through concrete ideas.

EXAMPLE

I have 6 pencils, and I want to divide them equally among 3 students. How many pencils will each student get?

(Answer: 6 ÷ 3 = 2)

When a student answers correctly, ask how the student arrived at that answer.

STUDENT PAGE
30
Division

Division Using Sets

Draw Xs on the board and lead students through discovering division concepts.

EXAMPLE

Outline groups of 6.

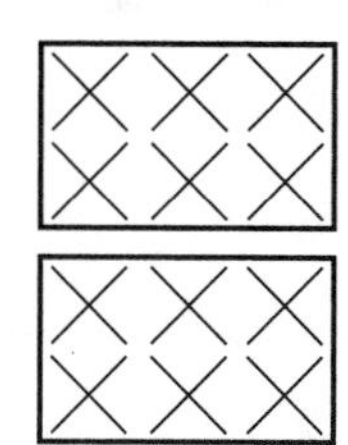

Ask:

How many Xs altogether?

How many Xs in each group?

How many groups are there?

Summarize: $12 \div 6 = 2$

Have students draw Xs and make up problems for each other.

STUDENT PAGE
30
Division

Using Manipulatives

Use manipulatives such as clothespins or game chips. Have students divide manipulatives into equal-sized groups and ask each other the same questions as in "Division Using Sets" above.

STUDENT PAGE
31
Division

Division Using Arrays

Arrays are arrangements of objects or symbols in orderly rows and columns. They can be very convincing models for illustrating division.

Ask:

How many dots in all? (Answer: 12)

How many in each group? (Answer: 4)

12 dots can be divided into how many groups of 4? (Answer: 3)

Summarize: $12 \div 4 = 3$

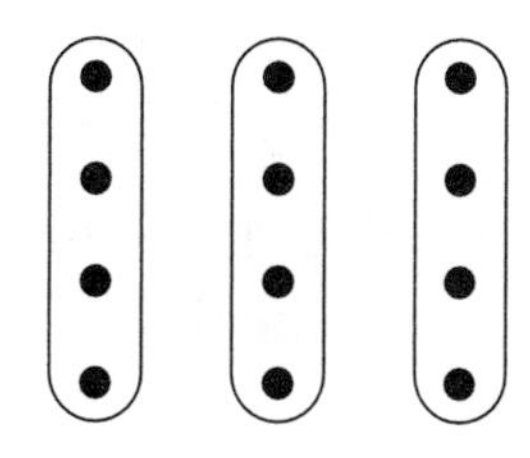

Provide many opportunities for students to make up their own problems with arrays.

Alternate Activity

Put out a quantity of manipulatives and have students use division facts to visualize division relationships.

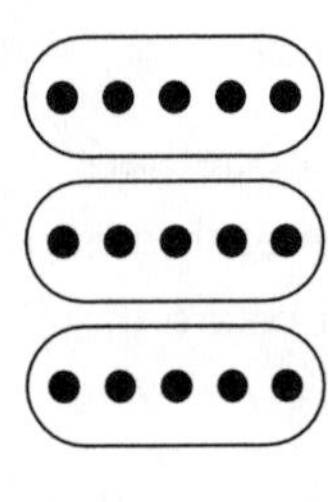

$15 \div 3 = 5$

$15 \div 5 = 3$

STUDENT PAGE
31
Division

Division Using a Number Line

Use the number line to illustrate the concept of division.

EXAMPLE $12 \div 4 = 3$ There are 3 groups of 4.

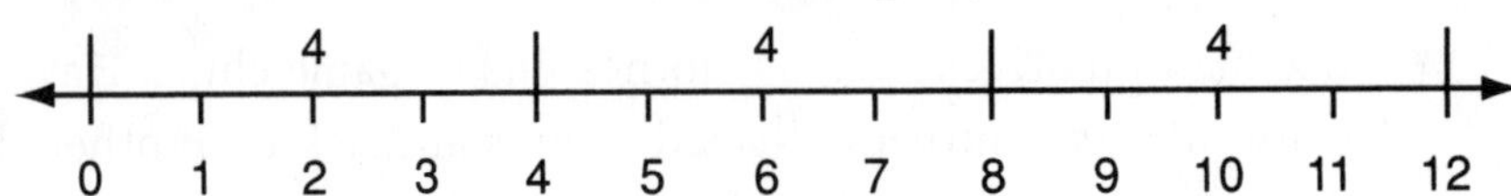

STUDENT PAGE
32
Division

Relating Multiplication and Division

Division is the **inverse** of multiplication, so have students think in terms of a **missing factor.**

EXAMPLES

$4 \times \square = 12$ OR $\begin{array}{r}\square\\4\overline{)12}\end{array}$ OR $\frac{12}{4} = \square$ OR $\frac{12}{\square} = 4$

It is important that students identify division when it is written in any of the different forms. Write multiplication facts on the board. Have students rewrite them in different forms as division problems.

Alternate Activities

1 Write problems in this form and ask students to determine what number the box represents.

EXAMPLES

$\begin{array}{r}\square\\6\overline{)42}\end{array}$ $\begin{array}{r}9\\4\overline{)\square\,\square}\end{array}$ $\begin{array}{r}6\\\square\overline{)24}\end{array}$ $\begin{array}{r}\square\\7\overline{)63}\end{array}$

2 Write problems in this form and ask students to determine what number the box represents.

EXAMPLES

A $\begin{array}{r} 6 \\ \times\ 8 \\ \hline \square \end{array}$ **B** $\begin{array}{r} \square \\ \times\ 5 \\ \hline 35 \end{array}$ **C** $\begin{array}{r} 4 \\ \times\ \square \\ \hline 12 \end{array}$

D $7 \times 6 = \square$ **E** $\square \times 7 = 49$ **F** $9 \times \square = 36$

3 List three numbers on the board and have students fill in each number sentence using only those numbers.

EXAMPLE [9] [63] [7]

$\square = \square \times 7$

$\square = 9 \times \square$

$\square = \square \div 7$

$\square = \square \div 9$

4 For each multiplication fact, have students write two division facts.

EXAMPLES

A $4 \times 7 = 28$

$28 \div 4 = 7$

$28 \div 7 = 4$

B $9 \times 8 = 72$

$72 \div 9 = 8$

$72 \div 8 = 9$

5 To review product-factor relationships, use this drill.

EXAMPLES

	Factor	Factor	Product
A	3	7	____
B	8	9	____
C	5	____	35
D	____	4	28
E	15	____	90

STUDENT PAGE
32
Long Division

Checking Division

Checking the solution helps students avoid errors and develop an ability to recognize reasonable answers.

EXAMPLES

A
```
   63
7)441
  42
   21
   21
    0
```

To check Example a, multiply 63 by 7.

The product is 441.

B
```
    12 R8
15)188
   15
    38
    30
     8
```

To check Example b, multiply 12 by 15 and add 8.

The product is 188.

STUDENT PAGE
32
Division

Estimate the Answer

This activity helps students focus on the correct placement of zeros in the quotient. Write problems like the ones below and have students circle the best estimate for each problem.

EXAMPLES

365 ÷ 70 =	5	50	500
639 ÷ 50 =	10	100	1,000
9,856 ÷ 48 =	20	200	2,000
15,275 ÷ 39 =	40	400	4,000
30,596 ÷ 15 =	20	200	2,000

STUDENT PAGE
56
Division

Relate the Operations

Replace the boxes with number to make each a true statement.

□ + 6 = 13	7 × □ = 42
8 + □ = 19	□ × 5 = 35
15 − □ = 7	63 ÷ □ = 7
□ − 4 = 10	□ ÷ 6 = 4

STUDENT PAGE
56
Division

Operation Strategies

1 Using a set of cards numbered 0–9 or a spinner, students take turns selecting a number at random.

2 On each turn, students should put one number in any box in a designated format. All students use the same format per game.

3 The object of the game is to build the largest possible number. The level of difficulty can be adjusted by using more complex formats.

EXAMPLE FORMATS

A (□ × □) − □

B (□ + □) ÷ □

C (□ × □) + (□ × □)

D (□□ + □) × (□□ − □)

STUDENT PAGE
56
Division

Multiple Operation Bingo

This version of bingo will help students learn to do mental math with multiple operations.

Have students draw a bingo board by sketching a 3 × 3 square.

1 Tell students to place at random any 9 numbers between 1 and 15 in the squares.

2 Make up ahead of time, for yourself, a set of problems with multiple operations.

3 Say each set of numbers and operations slowly, and let students compute mentally. When they figure an answer that is on the board, they cover that number.

4 Play continues until a player gets a bingo—any row, column, or diagonal covered.

EXAMPLES

A $(8 + 6) \div 2 \times 1 \div 7 = 1$

B $(15 \div 3) \times 5 \div 25 + 1 = 2$

C $(7 \times 7) + 7 \div 8 + 8 = 15$

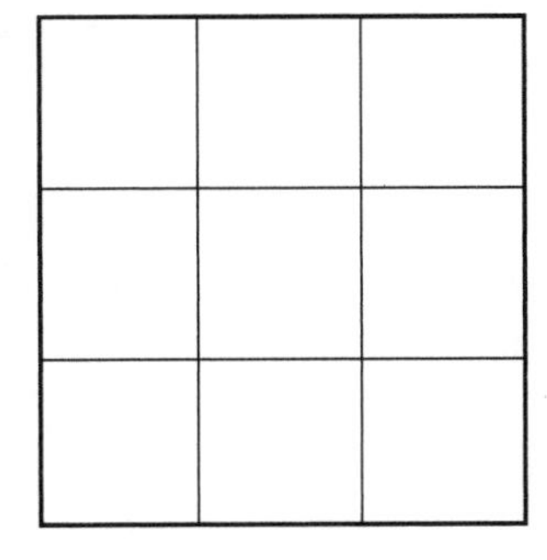

STUDENT PAGE
56
Division

Using Reference Points

Draw a diagram on the board or an overhead projector and let students decide which operations to perform to complete the boxes. Model several of these diagrams before the students work with them on their own. This exercise also lends itself well to cooperation among students working together in teams.

EXAMPLE

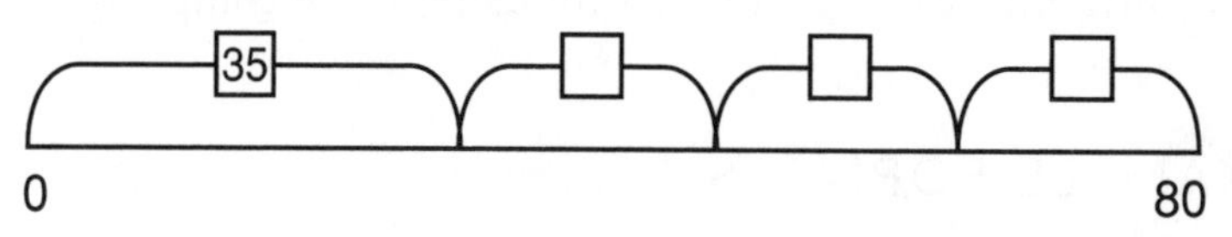

NOTE: These smaller reference points are always equidistant from each other.

Page 4: Pretest

1. 532
2. 1,596
3. 5,056
4. 18,525
5. 630
6. 11,270
7. 34,048
8. 92
9. 107
10. 1,461 R3
11. 46
12. 708
13. 270 R3
14. 456
15. 19 payments
16. $375
17. 224 miles
18. $625
19. 30 months
20. 240 hours

Pretest Evaluation Chart

If a student misses any problems in a skill area, assign the practice pages for that skill. However, students may need to complete all practice pages to reinforce areas of weakness.

Skill Area	Pretest Problem #	Skill Section	Review Page
Multiplication	1, 2, 3, 4, 5, 6, 7	7–23	24
Division	8, 9, 10, 11, 12, 13, 14	30–54	37, 55
Multiplication Problem Solving	16, 17, 20	25–28 62–68	29 69
Division Problem Solving	15, 18, 19	57–60 62–68	61 69
Life-Skills Math	All	70–73	74

Page 7: Meaning of Multiplication

1. 5
2. 4
3. 20
4. $5 \times 4 = 20$

Page 8: A Shortcut

1. $5 + 5 + 5 = 15$
   $3 \times 5 = 15$
2. 21, or $3 \times 7 = 21$
3. 16, or $4 \times 4 = 16$
4. 36, or $6 \times 6 = 36$

Page 9: Understanding Multiplication

1. 4
2. 6
3. 24
4. $4 \times 6 = 24$
5. $4 \times 3 = 12$
6. $5 \times 2 = 10$
7. $2 \times 8 = 16$
8. $4 \times 7 = 28$
9. $1 \times 6 = 6$
10. $6 \times 4 = 24$

Page 10: Special Multiplication Rules

1. a) 4×9 is the same as 9×4
   b) 5×7 is the same as 7×5
2. a) $1 \times 4 = 4$
   b) $7 \times 1 = 7$
   c) $1 \times 13 = 13$
   d) $9 \times 1 = 9$
3. a) $0 \times 5 = 0$
   b) $8 \times 0 = 0$
   c) $16 \times 0 = 0$
   d) $0 \times 35 = 0$

Page 11: Memorize Multiplication Facts

1. 4
2. 6
3. 8
4. 10
5. 12
6. 14
7. 16
8. 18
9. 9
10. 12
11. 15
12. 18
13. 21
14. 24
15. 27
16. 16
17. 20
18. 24
19. 28
20. 32
21. 36
22. 25
23. 30
24. 35
25. 40
26. 45
27. 36
28. 42
29. 48
30. 54
31. 49
32. 56
33. 63
34. 64
35. 72
36. 81

Page 12: Timed Multiplication Drill

1. 9
2. 64
3. 56
4. 35
5. 20
6. 30
7. 8
8. 14
9. 18
10. 25
11. 36
12. 16
13. 36
14. 81
15. 42
16. 7
17. 4
18. 0
19. 16
20. 28
21. 45
22. 48
23. 24
24. 5
25. 63
26. 42
27. 15
28. 12
29. 28
30. 32
31. 27
32. 6
33. 54
34. 21
35. 24
36. 21
37. 8
38. 12
39. 48
40. 18
41. 10
42. 0
43. 1
44. 54
45. 0
46. 56
47. 45
48. 72
49. 32
50. 63
51. 36
52. 72
53. 3
54. 40
55. 8
56. 24

Page 13: Practice Helps

1. 14
2. 45
3. 0
4. 64
5. 40
6. 6
7. 10
8. 24
9. 9
10. 49
11. 30
12. 54
13. 72
14. 18
15. 16
16. 81
17. 42
18. 27
19. 35
20. 16
21. 36
22. 21
23. 63
24. 36
25. 12
26. 56
27. 4
28. 20
29. 10
30. 9
31. 30
32. 35
33. 48
34. 8
35. 25
36. 54
37. 0
38. 18
39. 27
40. 24
41. 32
42. 28
43. 15
44. 21
45. 12
46. 40
47. 24
48. 12
49. 18
50. 12
51. 8
52. 2
53. 20
54. 3
55. 24
56. 72
57. 0
58. 40
59. 7
60. 8
61. 1
62. 15
63. 48
64. 28
65. 32
66. 18
67. 56
68. 42
69. 16
70. 6
71. 5
72. 0
73. 45
74. 24
75. 0
76. 4
77. 36
78. 15
79. 0
80. 63

Page 14: Find the Missing Numbers

1. 7
2. 3
3. 8
4. 8
5. 1
6. 8
7. 8
8. 5
9. 5
10. 5
11. 3
12. 5
13. 9
14. 7
15. 7
16. 2
17. 4
18. 9
19. 4
20. 0
21. 6
22. 1
23. 9
24. 7
25. 7
26. 3
27. 1
28. 1
29. 4
30. 5
31. 3
32. 8
33. 6
34. 6
35. 6
36. 3
37. 3
38. 2
39. 9
40. 5
41. 5
42. 0
43. 9
44. 9
45. 2
46. 1
47. 9
48. 6
49. 3
50. 2
51. 4
52. 0
53. 5
54. 4
55. 6
56. 2
57. 3
58. 5
59. 4
60. 3
61. 7
62. 7
63. 8
64. 9
65. 6
66. 0
67. 2
68. 9
69. 4
70. 1
71. 6
72. 7
73. 5
74. 4
75. 9
76. 6
77. 9
78. 2
79. 4
80. 9

Page 15: Multiplying by One-Digit Numbers

1. 48 **2.** 90 **3.** 68 **4.** 28 **5.** 86 **6.** 88 **7.** 80 **8.** 48 **9.** 568 **10.** 219 **11.** 248 **12.** 159 **13.** 129 **14.** 255 **15.** 156 **16.** 279

Page 16: Multiply and Regroup

1. 120 **2.** 132 **3.** 368 **4.** 329 **5.** 399 **6.** 558 **7.** 588 **8.** 531 **9.** 624 **10.** 315 **11.** 288 **12.** 268

Page 17: Hundreds, Tens, and Ones

1. 274 **2.** 987 **3.** 624 **4.** 525 **5.** 981 **6.** 896 **7.** 498 **8.** 892 **9.** 898 **10.** 872 **11.** 618 **12.** 678

Page 18: Regrouping Twice

1. 1,172 **2.** 3,970 **3.** 3,366 **4.** 4,476 **5.** 2,944 **6.** 3,283 **7.** 6,606 **8.** 772 **9.** 1,812 **10.** 3,810 **11.** 5,872 **12.** 768

Page 19: Multiplying by Two-Digit Numbers

1. 23 × 31 = 23 + 690 = 713

2. 12 × 11 = 12 + 120 = 132

3. 13 × 23 = 39 + 260 = 299

4. 53 × 11 = 53 + 530 = 583

5. 169

6. 1,089

7. 43 × 12 = 86 + 430 = 516

8. 528

9. 882

Page 20: Multiplying by Ones and Tens

1. 1,260

2. (3) 65 × 74 = 260 + 4550 = 4,810

3. 4,655 **4.** 2,242 **5.** 1,664 **6.** 5,628

Page 21: Multiplying Larger Numbers

1. (5)(6) 368 × 87 = 2576 + 29440 = 32,016

2. 41,952

3. (2)(1) 574 × 49 = 5166 + 22960 = 28,126

4. 19,142

5. (1)(1) 823 × 65 = 4115 + 49380 = 53,495

6. 30,672

Page 22: Multiplying by Multiples of Ten

1. 4,960 **2.** 1,410 **3.** 560 **4.** 5,950 **5.** 4,440 **6.** 2,720 **7.** 8,730 **8.** 4,400 **9.** 4,720 **10.** 31,560 **11.** 19,720 **12.** 15,060

Page 23: Larger Number on Top

Step 2 868 × 92 = 1736 + 78120 = 79,856

1. 2,310 **2.** 13,608 **3.** 64,542 **4.** 41,736 **5.** 6,344 **6.** 17,480

Page 24: Multiplication Review

1. a) 3 b) 6 c) 18 d) $3 \times 6 = 18$
2. 10, or $5 \times 2 = 10$
3. $6 \times 5 = 30$
4. 8
5. 0
6. 25
7. 9
8. 108
9. 270
10. 948
11. 567
12. 231
13. 1,978
14. 25,476
15. 43,500
16. 6,125

Page 25: Writing Number Sentences

1. $7 \times 4 = 28$
2. $3 × 7 = $21
3. $2 × 6 = $12
4. $15 \times 4 = 60$
5. $5 \times 9 = 45$
6. $7 \times 9 = 63$
7. $875 × 5 = $4,375
8. 8 × $5 = $40

Page 26: Writing Labels in Answers

1. 21 × 7 = 147 inches
2. 250 × 3 = 750 miles
3. 14 × 3 = 42 dollars
4. 75 × 7 = 525 strands of hair
5. 16 × 3 = 48 ounces
6. 9 × 28 = 252 pictures

Page 27: Practice Problem Solving

1. $45
2. 175 miles
3. 875 donuts
4. 312 pictures
5. $160
6. 144 eggs
7. 64 slices
8. 150 diapers
9. 72 stops
10. 42 pages

Page 28: Use the Correct Numbers

1. 200 chairs
   Extra information: 183 tickets
2. $225
   Extra information: 20 shorts at $10 each
3. 24 toys
   Extra information: 4 orphanages
4. 20 sticks of butter
   Extra information: 5 cups of frosting
5. $35
   Extra information: $7 original price
6. 63 tulips
   Extra information: 150 flowers

Page 29: Problem-Solving Review

1. 180 chairs
2. $40
3. 126 footballs
4. $40
5. $5
   Extra information: 10 mangos
6. $10
   Extra information: 1 can of soda
7. $425
   Extra information: 30 hours
8. 30 kilometers
   Extra information: 45 minutes per day

Page 30: Meaning of Division

A. 5
B. 5
1. 12
2. 2
3. 6
4. $12 \div 2 = 6$

Page 31: Division Readiness

1. 20
2. 4
3. 5
4. $20 \div 5 = 4$
5. $36 \div 9 = 4$
6. $30 \div 5 = 6$
7. $18 \div 2 = 9$
8. $36 \div 4 = 9$

Page 32: Relating Multiplication and Division

1. $7 \times 9 = 63$ is the opposite of $63 \div 9 = 7$
2. $5 \times 9 = 45$ is the opposite of $45 \div 5 = 9$
3. $7 \times 8 = 56$ is the opposite of $56 \div 7 = 8$
4. $5 \times 4 = 20$ is the opposite of $20 \div 4 = 5$
5. $3 \times 8 = 24$ is the opposite of $24 \div 8 = 3$
6. $36 \div 6 = 6$ is the opposite of $6 \times 6 = 36$
7. $27 \div 3 = 9$ is the opposite of $3 \times 9 = 27$
8. $49 \div 7 = 7$ is the opposite of $7 \times 7 = 49$
9. $72 \div 8 = 9$ is the opposite of $8 \times 9 = 72$
10. $18 \div 3 = 6$ is the opposite of $6 \times 3 = 18$

Page 33: Division Facts

A. $24 \div 4 = 6$
B. $24 \div 6 = 4$
C. $56 \div 8 = 7$
D. $56 \div 7 = 8$

1. 2
2. 3
3. 4
4. 5
5. 6
6. 7
7. 8
8. 9
9. 3
10. 4
11. 5
12. 6
13. 7
14. 8
15. 9
16. 4
17. 5
18. 6
19. 7
20. 8
21. 9
22. 5
23. 6
24. 7
25. 8
26. 9
27. 6
28. 7
29. 8
30. 9
31. 7
32. 8
33. 9
34. 8
35. 9
36. 9

Page 34: Practice Helps

1. 6
2. 4
3. 9
4. 4
5. 6
6. 7
7. 7
8. 7
9. 5
10. 3
11. 8
12. 6
13. 7
14. 8
15. 9
16. 6
17. 7
18. 6
19. 8
20. 9
21. 9
22. 9
23. 8
24. 5
25. 7
26. 3
27. 9
28. 7
29. 2
30. 3
31. 4
32. 8
33. 4
34. 7
35. 6
36. 5
37. 4
38. 3
39. 0
40. 3
41. 4
42. 5
43. 3
44. 1
45. 5
46. 5
47. 8
48. 5
49. 8
50. 6
51. 4
52. 5
53. 2
54. 3
55. 6
56. 9
57. 2
58. 2
59. 4
60. 2
61. 2
62. 1
63. 2
64. 8
65. 3
66. 1
67. 5
68. 2
69. 9
70. 0
71. 6
72. 1
73. 4
74. 3
75. 11
76. 20
77. 7
78. 6
79. 0
80. 1

Page 35: Two Ways to Show Division

1. $\begin{array}{r} 2 \\ 8\overline{)16} \end{array}$

2. $\begin{array}{r} 5 \\ 3\overline{)15} \end{array}$

3. $\begin{array}{r} 8 \\ 8\overline{)64} \end{array}$

4. $\begin{array}{r} 4 \\ 7\overline{)28} \end{array}$

5. $\begin{array}{r} 5 \\ 9\overline{)45} \end{array}$

6. $\begin{array}{r} 7 \\ 1\overline{)\ 7} \end{array}$

7. $\begin{array}{r} 4 \\ 10\overline{)40} \end{array}$

8. $\begin{array}{r} 7 \\ 8\overline{)56} \end{array}$

9. $\begin{array}{r} 3 \\ 6\overline{)18} \end{array}$

10. $\begin{array}{r} 5 \\ 5\overline{)25} \end{array}$

11. $\begin{array}{r} 1 \\ 9\overline{)\ 9} \end{array}$

12. $\begin{array}{r} 9 \\ 6\overline{)54} \end{array}$

13. $\begin{array}{r} 2 \\ 7\overline{)14} \end{array}$

14. $\begin{array}{r} 6 \\ 8\overline{)48} \end{array}$

15. $\begin{array}{r} 8 \\ 9\overline{)72} \end{array}$

16. $\begin{array}{r} 7 \\ 6\overline{)42} \end{array}$

Page 36: Basic Division

A. 2
B. 1
C. 2 R1
D. 2 R1
1. 2 R3
2. 2 R1
3. 6 R1
4. 2 R4

Think:
8 × 1 = 8
8 × 2 = 16 ✔
8 × 3 = 24

5. 1 R1
6. 3 R4
7. 2 R2
8. 4 R2
9. 5 R2
10. 1 R2
11. 3 R1
12. 2 R3

Page 37: Division Review

1. 3
2. 3
3. 8
4. 3
5. 2
6. 3 R2
7. 2 R3
8. 2 R6
9. 7 R7
10. 5 R2

Page 38: Steps for One-Digit Long Division

Step 5

```
   34
6)204
  18
   24
   24
    0
```

Long division has 5 steps. They are Step 1 **divide**, Step 2 **multiply**, Step 3 **subtract**, Step 4 **compare**, and Step 5 **bring down**.

Page 39: Practice the Steps

1.
```
   49
7)343
  28
   63
   63
    0
```

2.
```
   17
9)153
   9
   63
   63
    0
```

3.
```
   24 R4
6)148
  12
   28
   24
    4
```

4.
```
   37
4)148
  12
   28
   28
    0
```

5.
```
  27 R1
2)55
  4
  15
  14
   1
```

6.
```
   91 R1
8)729
  72
   09
    8
    1
```

7.
```
  22 R2
4)90
  8
  10
   8
   2
```

8.
```
   76
6)456
  42
   36
   36
    0
```

9.
```
   59
5)295
  25
   45
   45
    0
```

10.
```
   31 R2
8)250
  24
   10
    8
    2
```

11.
```
   87 R1
7)610
  56
   50
   49
    1
```

12.
```
   83
3)249
  24
   09
    9
    0
```

Page 40: Where to Start?

1.
```
   61
7)427
  42
   07
    7
    0
```

2.
```
  434 R1
2)869
  8
  06
   6
   09
    8
    1
```

3.
```
  192
4)768
  4
  36
  36
   08
    8
    0
```

4.
```
   72
9)648
  63
   18
   18
    0
```

5.
```
  211
4)844
  8
  04
   4
   04
    4
    0
```

6.
```
  131
4)524
  4
  12
  12
   04
    4
    0
```

Page 41: Zeros in the Answer

1. 4)816 = 204
816 − 8 = 016; 16 − 16 = 0

2. 3)270 = 90
270 − 27 = 00; 0 − 0 = 0

3. 7)753 = 107 R4
753 − 7 = 053; 53 − 49 = 4

4. 80 R6

5. 20

6. 109 R4

7. 180 R4

8. 109

9. 407 R1

Page 42: Using a Grid

1. 6)1,848 = 308
18; 048; 48; 0

2. 5)4,378 = 875 R3
40; 37; 35; 28; 25; 3

3. 9)5,283 = 587
45; 78; 72; 63; 63; 0

4. 3)2,169 = 723
21; 06; 6; 09; 9; 0

5. 6)8,755 = 1459 R1
6; 27; 24; 35; 30; 55; 54; 1

6. 7)5,621 = 803
56; 02; 0; 21; 21; 0

7. 5)9,845 = 1969
5; 48; 45; 34; 30; 45; 45; 0

8. 8)8,347 = 1043 R3
8; 03; 0; 34; 32; 27; 24; 3

9. 9)8,055 = 895
72; 85; 81; 45; 45; 0

Page 43: Dividing by One Digit

1. 1,200
2. 1,681
3. 1,151
4. 305
5. 254
6. 304
7. 479
8. 406
9. 190
10. 1,253
11. 909
12. 450

Page 44: Steps for Two-Digit Division

Step 5 90)2468 = 27 R38
2468 − 180 = 668; 668 − 630 = 38

Page 45: Dividing by Multiples of Ten

1. $40\overline{)346}$ = 8 R26
320
26

2. $20\overline{)196}$ = 9 R16
180
16

3. $10\overline{)58}$ = 5 R8
50
8

4. $90\overline{)248}$ = 2 R68
180
68

> Think:
> $90 \times 2 = 180$
> $90 \times 3 = 270$

5. $50\overline{)359}$ = 7 R9
350
9

> Think:
> $50 \times 6 = 300$
> $50 \times 7 = 350$

6. $70\overline{)500}$ = 7 R10
490
10

7. $80\overline{)230}$ = 2 R70
160
70

8. $50\overline{)94}$ = 1 R44
50
44

9. $40\overline{)143}$ = 3 R23
120
23

10. $30\overline{)99}$ = 3 R9
90
9

Page 46: Estimating Is Key

1. 70
2. 90
3. 50
4. 40
5. 20
6. 20
7. Estimate: 90 into 207 → 2
8. Estimate: 90 into 330 → 3
9. Estimate: 70 into 496 → 7
10. Estimate: 60 into 363 → 6

Page 47: Estimate the Quotient

1. $31\overline{)52}$ = 1
30 31 ☑ <52
21 ☑ <31

2. $32\overline{)76}$ = 2
30 64 ☑ <76
12 ☑ <32

3. $29\overline{)190}$ = 6
30 174 ☑ <190
16 ☑ <29

4. $81\overline{)526}$ = 6
80 486 ☑ <526
40 ☑ <81

5. $58\overline{)562}$ = 9
60 522 ☑ <562
40 ☑ <58

6. $48\overline{)312}$ = 6
50 288 ☑ <312
24 ☑ <48

7. $18\overline{)63}$ = 3
20 54 ☑ <63
9 ☑ <18

8. $79\overline{)386}$ = 4
80 316 ☑ <386
70 ☑ <79

9. $72\overline{)266}$ = 3
70 216 ☑ <266
50 ☑ <72

Page 48: Adjust When Necessary

1. 38)268 — quotient ~~6~~ 7
   266 ☑
   2 ☑

2. 33)159 — quotient ~~5~~ 4
   132 ☑
   27 ☑

3. 29)209 — quotient 7
   203 ☑
   6 ☑

4. 89)713 — quotient ~~9~~ 8
   712 ☑
   1 ☑

5. 67)275 — quotient 4
   268 ☑
   7 ☑

6. 33)159 — quotient ~~5~~ 4
   132 ☑
   27 ☑

7. 27)172 — quotient 6
   162 ☑
   10 ☑

8. 33)198 — quotient ~~5~~ 6
   198 ☑
   0 ☑

9. 92)366 — quotient ~~4~~ 3
   276 ☑
   90 ☑

Page 49: Decide Where to Start

1. 41)307 — XX□
2. 36)621 — X□□
3. 45)443 — XX□
4. 62)621 — X□□
5. 14)286 — X□□
6. 4)832 — □□□
7. 16)174 — X□□
8. 91)103 — XX□
9. 7)708 — □□□
10. 83)868 — X□□

Page 50: Round to Divide

1. 18)49 — quotient 2
   [20] 36 ☑
   13 ☑

2. 47)230 — quotient 4
   [50] 188 ☑
   42 ☑

3. 53)263 — quotient 4
   [50] 212 ☑
   51 ☑

4. 88)326 — quotient 3
   [90] 264 ☑
   62 ☑

5. 54)395 — quotient 7
   [50] 378 ☑
   17 ☑

6. 26)81 — quotient 3
   [30] 78 ☑
   3 ☑

7. 19)168 — quotient 8
   [20] 152 ☑
   16 ☑

8. 34)128 — quotient 3
   [30] 102 ☑
   26 ☑

9. 59)472 — quotient 8
   472 ☑
   0 ☑

10. 67)297 — quotient 4
    268 ☑
    29 ☑

11. 73)432 — quotient 5
    365 ☑
    67 ☑

12. 23)198 — quotient 8
    184 ☑
    14 ☑

Page 51: Digits in the Quotient

1. 63)9,253 — X □□□
2. 9)68,259 — X□ □□□
3. 189)2,237 — X X□□
4. 98)1,000 — X X□□
5. 602)16,856 — XX X□□
6. 213)943 — XX□
7. 90)800 — XX□
8. 743)1,649 — X XX□
9. 5)4,096 — X □□□
10. 73)8,506 — X □□□
11. 8)596 — X□□
12. 43)2,476 — X X□□
13. 49)39,741 — XX □□□
14. 19)37,520 — X□ □□□
15. 973)6,190 — X XX□

Page 52: Estimate the Answer

1. 7
2. 2,000
3. 30
4. 700
5. 9
6. 1,000
7. 300
8. 40

Page 53: Divide by Two Digits

1.
```
      548
 72)39,456
    360
     345
     288
      576
      576
        0
```
2. 839
3. 16 R7
4. 782
5. 79 R1
6. 492

Page 54: Practice Helps

1.
```
     107
 23)2,461
    23
     16
      0
     161
     161
       0
```
2. 306
3. 408
4. 19
5. 217
6. 6 R19
7. 108 R9
8. 408
9. 903

Page 55: Division Review

1. $36 \div 6 = 6$
2. $7 \times 8 = 56$ is the opposite of $56 \div 8 = 7$
3.
```
   7
 9)63
```
4. 3 R2
5. 28
6. 21
7. 207
8. 701
9. 8 R16
10. 50
11. 8 R4; round to 80
12. 80 R20 round to 20
13. 40

Page 56: Putting It All Together

1. $120 < 126$
2. $528 < 588$
3. $92 = 92$
4. $8 > 7$
5. $2,304 = 2,304$
6. $473 > 453$
7. $78 < 81$
8. $4,224 > 4,053$
9. $99 < 100$
10. $658 < 703$
11. $72 = 72$
12. $38 > 36$

Page 57: Writing Number Sentences

1. $45 ÷ 3 = $15
2. 24 ÷ 3 = 8 hours
3. 8 ÷ 4 = 2 pieces
4. $12 ÷ 3 = $4
5. $63 ÷ 9 = $7
6. $74 ÷ 2 = $37

Page 58: Writing Labels in Answers

1. $175 ÷ $25 = 7 weeks
2. $57 ÷ 3 = 19 dollars
3. 36 ÷ 6 = 6 sacks
4. 440 ÷ 4 = 110 yards
5. 18 ÷ 2 = 9 calls; calls
6. $16 ÷ 4 = 4 dollars; dollars
7. $63 ÷ $21 = 3 shirts; shirts
8. $30 ÷ 3 = 10 dollars; dollars

Page 59: Think About the Remainder

1. 2 R4; 3 tables
2. 31 R1; 31 boards
3. 2 R1; 2 cakes
4. 3 R1; 4 Popsicles
5. 6 R10; 6 cars
6. 15 R2; 16 nickels

Page 60: Use the Correct Numbers

1. $9
   Extra information: 4 pizzas
2. $7
   Extra information: 5 weeks
3. 5 days
   Extra information: 50 yards
4. 2 tables
   Extra information: 2 pies
5. $25
   Extra information: $9 movie ticket
6. $5 per hour
   Extra information: 7 days

Page 61: Problem-Solving Review

1. 98 ÷ 49 = 2 hot dogs
2. $35 ÷ 7 = $5
3. $42 ÷ $7 = 6 weeks
4. $66 ÷ 2 = $33
5. 65 ÷ 9 = 7 R2; 2 tickets left over
6. 6 dimes
   Extra information: 2 cans of soda
7. $25
   Extra information: 2 concerts
8. $2
   Extra information: used 15 passes

Page 62: Mixed Multiplication and Division

1. 7 × 16 = 112 boxes
2. 24 ÷ 8 = 3 feet
3. 75 × 5 = 375 miles
4. 8¢ × 15 = $1.20
5. 135 ÷ 15 = 9 weeks
6. $95 × 3 = $285
7. 75 × 7 = 525 times
8. 285 ÷ 15 = 19 payments
9. $25 × 15 = $375
10. 270 ÷ 5 = 54 miles

Page 63: Choose the Operation

1. division
2. subtraction
3. multiplication
4. addition
5. ×
6. ×
7. +
8. –
9. ÷
10. +

Page 64: Write the Question

Answers will vary, but should represent a knowledge of number sentences.

Page 65: Decide the Operation

1. $25 + $17 = $42
2. $10 – $6 = $4
3. 115 ÷ 23 = 5 buses
4. 7 × 5 = 35 minutes
5. 485 – 138 = 347 miles
6. 28 ÷ 7 = 4 students
7. 500 ÷ 25 = 20 gallons
8. $18 × 5 = $90

Page 66: Mixed Practice

1. 485 + 1,340 = $1,825; 1,825
2. 18 × 9 = 162; 162
3. 9 × $20 = $180; 180
4. $215 – $119 = $96; 96
5. $450 + $15 = $465; 465
6. $765 ÷ 3 = $255; 255
7. 15 × 8 = 120; 120
8. 12 × 4 = 48; 48

Page 67: Looking at Word Problems

1. a) 175
   b) 175
   c) 175
   d) 175
   e) 175
2. a) 90
   b) 90
   c) 90
3. a) 5
   b) 5
   c) 5
   d) 5
   e) 5

Page 68: Apply Your Skills

1. D
2. B
3. A
4. C
5. 15
6. 15
7. 150 > 135
8. 135 < 150
9. 135 ≠ 150
10. \$135 – \$78 = \$57; \$57
11. \$45 ÷ 3 = \$15; \$15

Page 69: Mixed Problem-Solving Review

1. 315 ÷ 35 = 9
2. 55 × 4 = 220
3. 76 ÷ 19 = 4
4. \$2 × 18 = \$36
5. 59 + 75 = 134
6. \$136 ÷ 4 = \$34
7. Answers will vary.
8. a) 9
   b) 9
   c) 9
   d) 9

Page 70: Everyday Math

1. a) 6
   b) 12
2. \$125
3. a) 2 × 15 = \$30
   b) 3 × \$8 = \$24
   c) \$30 + \$24 = \$54
4. a) \$13
   b) 3 × \$13 = \$39
5. \$375
6. a) 4
   b) 4 × \$5 = \$20

Page 71: Picture Problems

1. 144 color prints
2. \$96
3. \$28
4. 384 miles per hour
5. \$6
6. 375 miles

Page 72: Weekly Pay

	Hours Worked	Paid
1.	7	\$63
2.	5	\$45
3.	4	\$36
4.	8	\$72
5.	6	\$54
6.	30	\$270

7. 30 hours
8. \$270
9. a) \$18
   b) \$18
   c) \$27
10. 2 hours
11. \$1,080
12. 26 weeks

Page 73: Weekly Sales

	Unit Price	Amount
1.	$260	$2,080
2.	$578	$2,890
3.	$53	$1,378
4.		$6,348

5. 13 months
6. 17 months
7. 51 hours
8. $6,348; $6,348 ÷ 2 = $3,174

Page 74: Life-Skills Math Review

1. $70
2. $4.25
3. $280
4. 23 payments
5. 48 cans
6. 108 apples
7. $45
8. 390 miles

Page 75: Cumulative Review

1. 3,225
2. 168
3. 40
4. 1
5. 280
6. 21
7. 48 R2
8. 497

Page 75: Posttest

1. 2,491
2. $83
3. 38,554
4. $120
5. 1,537
6. 459
7. 41,912
8. 407
9. $5,040
10. 16,020
11. 11 months
12. 83
13. 10,192
14. 175
15. 406 R5
16. $9
17. $180
18. 387
19. 544 students
20. 8 pounds

Posttest Evaluation Chart

If a student misses one or more problems in a skill area, assign a review of the practice pages for that skill.

Skill Area	Posttest Problem #	Skill Section	Review Page
Multiplication	1, 3, 5, 7, 10, 13	7–23	24
Division	6, 8, 12, 14, 15, 18	30–54	37, 55
Multiplication Problem Solving	4, 9, 17, 19	25–28 62–68	29 69
Division Problem Solving	2, 11, 16, 20	57–60 62–68	61 69
Life-Skills Math	All	70–73	74

DECIMALS
Addition & Subtraction

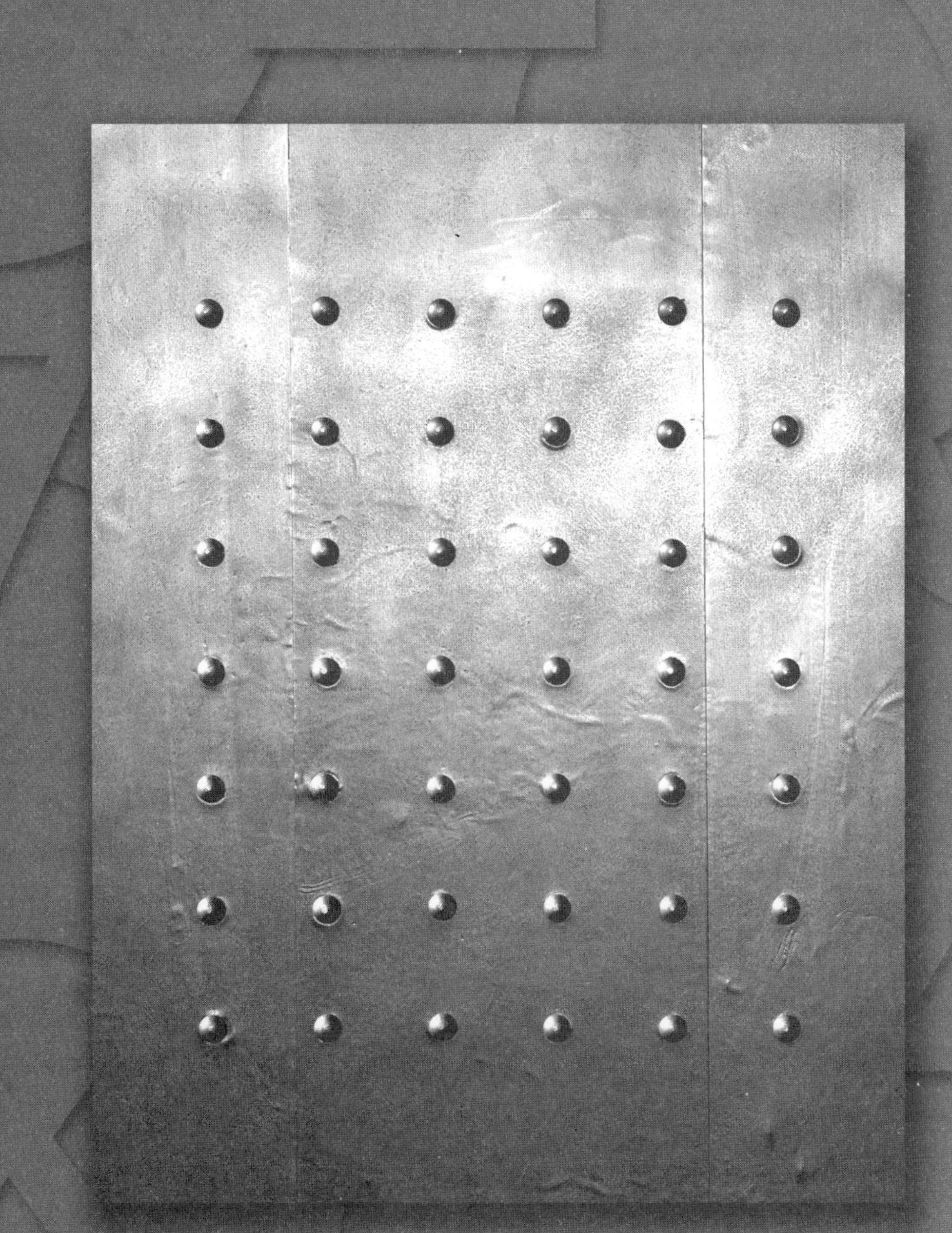

Traditionally, fractions have been introduced before decimals. However, this is changing due to the increasing emphasis on calculators, the application of metrics in measurement, and students' personal familiarity with decimals through our money system. Therefore, *Number Sense* has been designed so the teaching of decimals and the teaching of fractions are not dependent on each other. The teaching sequence can be easily interchanged.

Student Glossary

Acquainting students with definitions of key math terms and life-skills concepts will enhance their mastery of the materials. Below are words defined in the student text. A glossary is provided at the end of the student text and on page 243 of the *Teacher's Resource Guide and Answer Key*.

bring down
budget
checking account
circle
combined earnings
coupons
deposit
digit
discount coupon
expanded form
increased
late fee
line up
mixed decimal
number relation symbol
operation symbol
place value
placeholder
regroup (borrow)
regroup (carry)
regular price
sale price
savings account
tax
tip

STUDENT PAGE

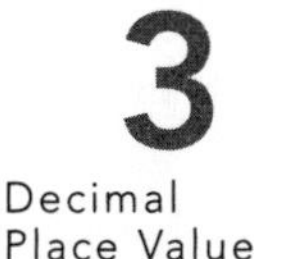

Decimal
Place Value

Name the Same Numbers

If fractions have been introduced before decimals, point out that there are two different ways of naming the same number: the fraction form and decimal form.

You can explain that decimals are also called **decimal fractions** and that tenths, hundredths, and thousandths can be written both as decimals and as fractions. Students should practice writing both forms of the same number.

EXAMPLES

Decimals to Fractions	Fraction to Decimals
$.9 = \frac{\square}{10}$	$\frac{2}{10} = .\_$
$.25 = \frac{\square}{100}$	$\frac{7}{100} = .\_\,\_$
$.019 = \frac{\square}{1{,}000}$	$\frac{231}{1{,}000} = .\_\,\_\,\_$

STUDENT PAGE

3

Decimal
Place Value

Using Money

Most students become familiar with money at an early age. It often helps to relate fractions and decimals using dollar amounts. Explain that since a dollar has 100 pennies, a dime (10 cents) is .1, or one-tenth, of a dollar or 10 out of 100 cents.

EXAMPLES

.1 or $\frac{1}{10}$ of one dollar is ______cents.

.2 or $\frac{2}{10}$ of one dollar is ______cents.

.5 or $\frac{5}{10}$ of one dollar is ______cents.

.8 or $\frac{8}{10}$ of one dollar is ______cents.

.4 or $\frac{4}{10}$ of one dollar is ______cents.

Alternate Activity

One dollar is equal to ______ pennies.

15 pennies are equal to ______ (decimal) or ______ (fraction) of one dollar.

75 pennies are equal to ______ (decimal) or ______ (fraction) of one dollar.

25 pennies are equal to ______ (decimal) or ______ (fraction) of one dollar.

6 pennies are equal to ______ (decimal) or ______ (fraction) of one dollar.

5 pennies are equal to ______ (decimal) or ______ (fraction) of one dollar.

4 pennies are equal to ______ (decimal) or ______ (fraction) of one dollar.

STUDENT PAGE 15
Decimal Place Value

Naming the Decimal

Have students complete a chart showing 1) whether the decimals are tenths, hundredths, or thousandths and 2) indicating the number of decimal places for each decimal.

EXAMPLE

Decimal	Tenths	Hundredths	Thousandths	Number of Decimal Places
.72		✓		2
9.3				
.134				

STUDENT PAGE 15
Decimal Place Value

Place-Value Patterns

Use a place-value chart to strengthen students' understanding of the place-value system.

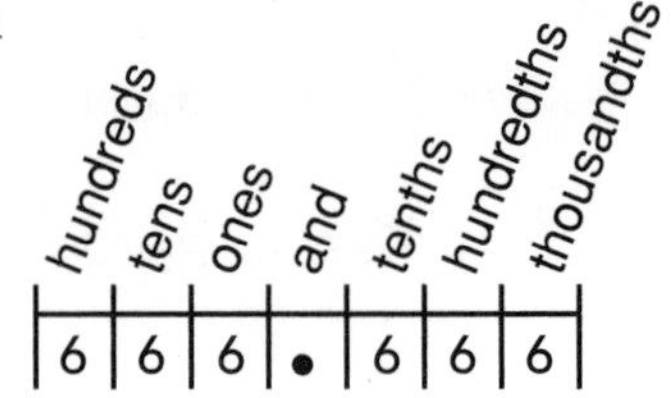

Start with the whole numbers from right to left.

1 The value of the tens place is ________ times the ones place.
(Answer: 10 60 is 10 times 6)

2 The value of the hundreds place is ________ times the tens place.
(Answer: 10 600 is 10 times 60)

Now work the other way with the whole numbers.

3 The value of the tens place is ______ of the hundredths place.
(Answer: $\frac{1}{10}$ 60 is $\frac{1}{10}$ of 600)

4 The value of the ones place is ______ of the hundredths place.
(Answer: $\frac{1}{10}$ 6 is $\frac{1}{10}$ of 60)

Finally, apply this patterning to decimals.

5 The value of the tenths place is ______ of the ones place.
(Answer: $\frac{1}{10}$ .6 is $\frac{1}{10}$ of 6)

6 The value of the hundredths place is ______ of the value of the tenths place.

By discovering this pattern of relationships, students will understand the structure of the decimal system.

STUDENT PAGE **15** Decimal Place Value

Write Decimals in Words

If students have worked with fractions, you can show them how decimals are a type of fraction. Write fractions and have students fill in the place-value words: *tenths, hundredths,* and *thousandths.*

EXAMPLES

$\frac{4}{100}$ = 4________ $\frac{7}{10}$ = 7________

$\frac{105}{1000}$ = 105________ $\frac{76}{100}$ = 76________

STUDENT PAGE **18** Decimal Place Value

Rewriting Decimals

To have students learn how to compare decimals, write a group of decimals on the board. Have the students rewrite the decimals so that each decimal has the same number of decimal places.

EXAMPLES

.05 .6 .159 = .050 .600 .159

.5 15 .007 .65 = ____ ____ ____ ____

7 3.5 .905 = ____ ____ ____

.93 4.8 7 .805 = ____ ____ ____ ____

Once decimals have the same number of decimal places, students can read them as whole numbers to compare.

EXAMPLES

.050 = 50 (thousandths)
.600 = 600 (thousandths)
.159 = 159 (thousandths)

STUDENT PAGE
24
Decimal Place Value

Using the Symbols

Have students use the symbol >, <, or = to make each statement true.

EXAMPLES

.45 ◯ .450

.9 ◯ .09

.370 ◯ .37

.8 ◯ .75

.022 ◯ .22

STUDENT PAGE
24
Addition

Build the Decimals

Write several digits and have students rearrange the digits to name the largest and smallest possible numbers.

EXAMPLE 6 7 2 9 3

Largest number ____ ____. ____ ____ ____ (Answer: 97.632)

Smallest number ____ ____. ____ ____ ____ (Answer: 23.679)

Alternate Activity

1. From a deck of cards marked 0–9, teams take turns choosing at random one card at a time. Using a format similar to the ones below, students write a number in any box they choose. Students should only use one format per game.
2. After each draw, the card is mixed in for the next draw.
3. Students can use this game to build either the largest possible decimal number or the smallest possible number.

EXAMPLES

Format A: □□□.□

Format B: □□.□□

Format C: □.□□□

Alternate Activity

Use the same digit several times to give students an idea of place value.

EXAMPLE

Write as many different decimal numbers as you can using the digits below:

3 3 3

Arrange the decimal numbers from largest to smallest.
(Answer: 333, 33.3, 3.33, and .333)

What is the largest number?
(Answer: 333)

What is the smallest number?
(Answer: .333)

STUDENT PAGE
28
Addition

Catalog Shopping

Bring in a mail-order catalog. Students can pretend to order five items from the catalog. The student who comes closest to $50.00, without going over, wins. Change the total to create new games.

STUDENT PAGE
30
Addition

Estimate the Answer

Have students circle the best estimated answer for each problem. First, show them how to round to the nearest whole number.

EXAMPLES

15.43	+	54.81	=	70	700	7,000
60.5	+	99.8	=	16	160	1,600
25.7	+	3.9	=	3.0	30	300
499.9	+	25.05	=	5.25	52.5	525

STUDENT PAGE
36
Addition
Problem
Solving

Rounding Money Values

Demonstrate how to round money values to the nearest dollar and to the nearest cent. Discuss when you might need to round to the nearest dollar. For example, estimating the cost of items while shopping. Discuss when you might need to round to the nearest cent. For example, estimating the price per gallon at a gas station.

EXAMPLES

Round to the nearest dollar.		Round to the nearest cent.	
$49.62	(Answer: $50)	$5.8475	(Answer: $5.85)
$3.16	(Answer: $3)	$26.067	(Answer: $26.07)
$14.50	(Answer: $15)	$13.506	(Answer: $13.51)
$.76	(Answer: $1)	$.932	(Answer: $.93)
$15.79	(Answer: $16)	$4.9381	(Answer: $4.94)
$29.40	(Answer: $29)	$9.196	(Answer: $9.20)

STUDENT PAGE **36** Addition Problem Solving

Estimate Dollar Total While Shopping

Ask students if $20 will be enough money to buy the following items? Encourage students to estimate.

______ $5.95	______ $2.10	______ $14.45
______ $6.25	______ $3.25	______ $2.50
______ $9.15	______ $12.85	______ $2.35

Alternate Activity

Using grocery ads, create a list of items including the prices. Have students estimate whether $50 will be enough money to buy all the items.

STUDENT PAGE **36** Addition Problem Solving

Money Sense

Make a grid similar to the one below. Have students decide the least number of coins possible to total each amount.

EXAMPLE

	Pennies	Nickels	Dimes	Quarters
$.17	2	1	1	
$.64				
$.23				

DECIMALS: Addition and Subtraction

STUDENT PAGE
37
Addition
Problem
Solving

Making Change

Present the idea of purchases where the change will be less than $1.00. Have students use real coins to make change. Instruct them to use the least number of coins possible.

EXAMPLES

Purchase is $4.15
Paid $5.00
Dimes ______
Quarters ______

Purchase is $.35
Paid $1.00
Nickels ______
Dimes ______
Quarters ______

Alternate Activity

Use amounts of money where the change will be 1) less than $5.00 and 2) less than $10.00.

STUDENT PAGE
37
Addition
Problem
Solving

Newspaper Mathematics

Have students bring in the newspaper on the day of the week that the grocery sales are advertised. Tell your students they have an imaginary $65.00 to purchase a week's groceries as advertised in the local paper. Have them make a shopping list and keep a record of what they spent.

STUDENT PAGE
38
Addition
Problem
Solving

Sketch Reference Points

Teaching students to sketch reference points is a good problem-solving strategy.

EXAMPLES

A The dinner costs $14.35. How much change will you get back from $20.00?

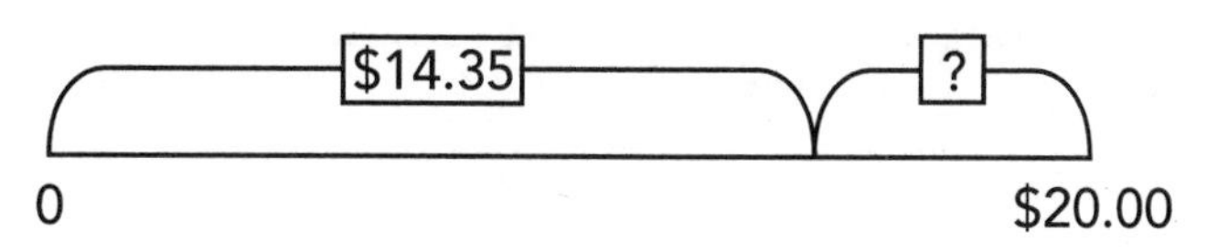

B Adam rode his bike 31.4 miles on Monday and 43 miles on Tuesday. How many miles did he ride in two days?

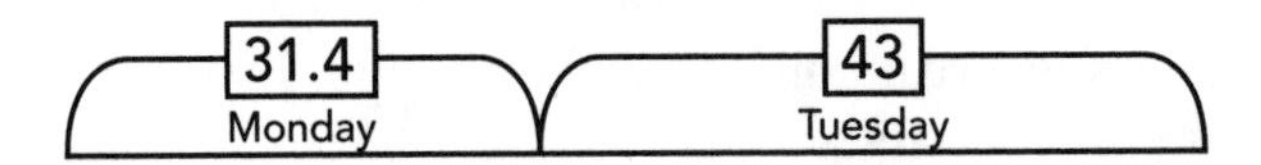

STUDENT PAGE
39
Addition Problem Solving

Problem Solving

Let your students become active participants in creating and solving real-life problems.

Have students

1. Fill in the blanks with reasonable numbers.
2. Write a question about the facts.
3. Complete a number sentence for each problem.
4. Write the answer in sentence form. This helps students clarify their thinking—"Does the answer make sense?"

EXAMPLE

Jodi bought a camera for $_____ and a calculator for $_____.

Question: ______________________________

_____ _____ _____ = _____

Number, Operation Symbol, Number, Answer

Answer: ____________

STUDENT PAGE
54
Subtraction

Relating Addition and Subtraction

Ask students to write number sentences and solve them.

EXAMPLES

What would you add to $23.45 to get $50.00?
$23.45 + _____ = $50.00

What would you add to $136.15 to get $225.75?
$136.15 + _____ = $225.75

Alternate Activities

1 Write an additional problem on the board and have your students write and solve two related subtraction problems.

Addition Problem	Related Subtraction Problems	
$6.94 + 3.09 = 10.03$	$10.03 - 3.09 = 6.94$	$10.03 - 6.94 = 3.09$

2 Have students find two decimals whose difference is 9.35. Answers will vary. Discuss how students arrived at their answers.

EXAMPLES

9.35 + [5.89] = [15.24] so 15.24 − 5.89 = 9.35

9.35 + [25.36] = [34.71] so 34.71 − 25.36 = 9.35

3 Have students find two decimals whose sum is 8.34. Answers will vary. Discuss how students arrived at their answers.

EXAMPLES

8.34 − [3.96] = [4.38] so 4.38 + 3.96 = 8.34

8.34 − [6.90] = [1.44] so 1.44 + 6.90 = 8.34

STUDENT PAGE
54
Subtraction

Does It Make Sense?

Demonstrate how students can easily detect if they have misplaced a decimal point or incorrectly lined up numbers. Tell them to check their answers by rounding to the nearest whole number. This way they can judge how reasonable their answers are.

EXAMPLE .84 + 2.9 + 8 =

aligned incorrectly	estimate to check	aligned correctly
.84	.84 rounds to 1	.84
2.9	2.9 rounds to 3	2.9
+ 8	8 rounds to 8	+ 8
3.82	12	11.74
wrong answer	answer should be close to 12	answer is close to 12

STUDENT PAGE
55
Subtraction

Find the Missing Value

Students enjoy making up problems to challenge their classmates. This kind of activity provides practice in adding and subtracting decimals. Draw a scale on the board. Explain that both sides must balance. Therefore, the total must be the same.

EXAMPLE

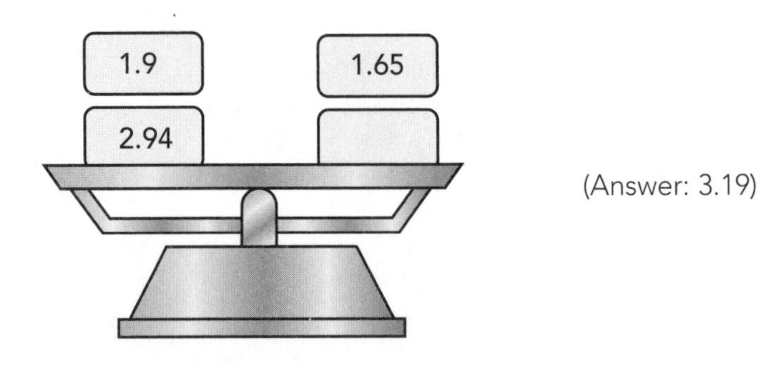

(Answer: 3.19)

Alternate Activity

Have one student fill in 3 out of the 4 numbers needed and then challenge other students to balance the scale.

STUDENT PAGE
55
Subtraction

Find the Sum and Difference

Using two numbers in boxes, have students write number sentences to find the sum and difference.

EXAMPLE

9.8 21.9 Answer: 9.8 + 21.9 = ______

Answer: 21.9 − 9.8 = ______

STUDENT PAGE
57
Subtraction
Problem
Solving

Listing Prices from the Newspaper

Using advertisements in a newspaper, have students make a list of ten items that are priced under $20.00. Then they can invent addition and subtraction problems using the list.

EXAMPLES

What is the difference in cost between the most expensive item and the least expensive item?

The five most expensive items cost how much?

If you buy the three least expensive items, how much change will you get back from $20.00?

Alternate Activity

Make a list of 3 items that are priced under $50, $100, $1,000, etc. Invent addition and subtraction problems using the list.

STUDENT PAGE
60
Subtraction
Problem
Solving

Newspaper Mathematics

Nearly every advertisement in newspapers shows how numbers play a vital role in real-life activities. Students can easily invent and illustrate problems using pictures.

Have students

1. Write a question about the facts.
2. Complete a number sentence for each problem.
3. Write the answer in sentence form. This helps students clarify their thinking—"Does the answer make sense?"

EXAMPLE

Question: ______________________________

Number Sentence: ______ ______ ______ = ______
(Operation Symbol under the second blank)

Answer: ______________________________

Was $125.65
Now $99.75

Alternate Activity

Assign students to create problems and illustrate them using pictures that have 1) more information than is needed to solve the problem and 2) not enough information to solve the problem. Give them models for each type of problem.

Page 4: Pretest

1. 4
2. 8.36
3. .420
4. 2.5
5. 21.27
6. 40.83
7. 106.10 or 106.1
8. 29.94
9. 286.50 or 286.5
10. 13.67
11. .22
12. 33.64
13. 17.74
14. 22.35
15. $7.47
16. $23.70
17. 86.3 miles
18. 4.4°
19. 20.25 inches
20. $.32 million

Pretest Evaluation Chart

If a student misses any problems in a skill area, assign the practice pages for that skill. However, students may need to complete all practice pages to reinforce areas of weakness.

Skill Area	Pretest Problem #	Skill Section	Review Page
Place Value	1, 2, 3, 4	7–24	25
Addition	5, 6, 7, 8, 9	26–34	35
Subtraction	10, 11, 12, 13, 14	45–55	56
Addition Word Problems	16, 17, 19	36–43	44
Subtraction Word Problems	15, 18, 20	57–66	67
Life-Skills Math	All	68–73	74

Page 7: Meaning of Tenths

1.

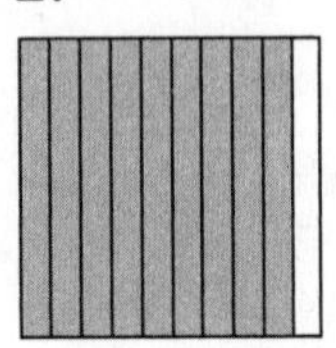

6 out of 10

2.

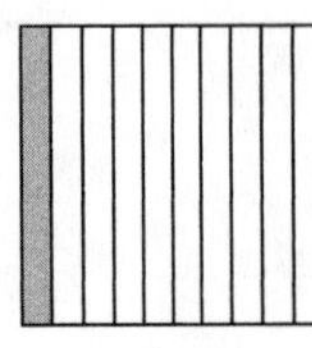

9 out of 10

3.

1 out of 10

4.

0 .2 .5 1

.1 one tenth

.3 three tenths

.4 four tenths

.7 seven tenths

.9 nine tenths

Page 8: Comparing Tenths

1. greater
2. less
3. less
4. less
5. greater
6. greater
7. greater
8. greater
9. smaller
10. larger

Page 9: Meaning of Hundredths

1.

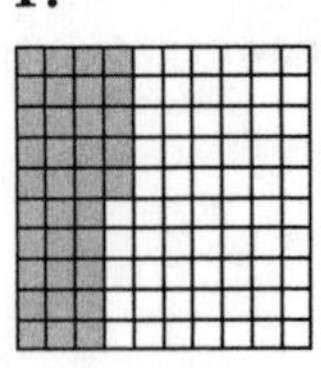

35 out of 100

2.

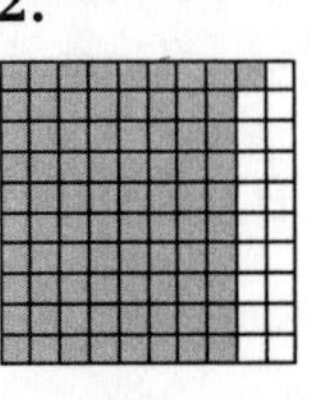

81 out of 100

3.

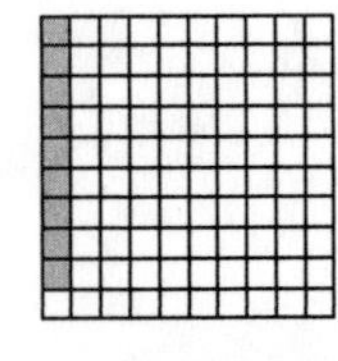

9 out of 100

4. 100
5. .15
6. .75
7. .25
8. .01
9. .50
10. .04

Page 10: Comparing Hundredths

1. greater
2. greater
3. greater
4. less
5. less
6. greater
7. less
8. less

Page 11: Meaning of Thousandths

1. .035
2. .069
3. .350
4. .936
5. .525
6. 250 out of 1,000 .250
7. 40 out of 1,000 .040
8. 485 out of 1,000 .485
9. 806 out of 1,000 .806
10. 15 out of 1,000 .015
11. 685 out of 1,000 .685
12. 167 out of 1,000 .167

Page 12: Comparing Thousandths

1. less
2. greater
3. greater
4. less
5. greater
6. greater
7. less
8. greater
9. greater
10. less

Page 13: Place–Value Readiness

	hundreds	tens	ones	and	tenths	hundredths	thousandths
A.	8	7	0	.	2	1	4

1. ones
2. hundredths
3. tenths
4. thousandths
5. tens
6. 0
7. 8
8. 7
9. 1
10. 2

Page 14: Place Values

1. a) 7 tens
   b) 1 one
   c) 6 tenths
2. a) 9 tens
   b) 4 ones
   c) 3 tenths
   d) 5 hundredths
3. a) 8 tens
   b) 9 ones
   c) 4 tenths
   d) 6 hundredths
   e) 3 thousandths

Page 15: Reading Decimals

	tens	ones	and	tenths	hundredths	thousandths	
1. 71.647	7	1	.	6	4	7	71 and 647 thousandths
2. 50.94	5	0	.	9	4		50 and 94 hundredths
3. 1.6		1	.	6			1 and 6 tenths
4. 16.345	1	6	.	3	4	5	16 and 345 thousandths
5. 3.7		3	.	7			3 and 7 tenths
6. 8.04		8	.	0	4		8 and 4 hundredths
7. 17.39	1	7	.	3	9		17 and 39 hundredths
8. 98.507	9	8	.	5	0	7	98 and 507 thousandths
9. 7.53		7	.	5	3		7 and 53 hundredths
10. 84.2	8	4	.	2			84 and 2 tenths
11. 4.008		4	.	0	0	8	4 and 8 thousandths
12. 7.45		7	.	4	5		7 and 45 hundredths
13. 6.19		6	.	1	9		6 and 19 hundredths
14. 29.133	2	9	.	1	3	3	29 and 133 thousandths
15. 2.4		2	.	4			2 and 4 tenths

Page 16: Write the Place Value

1. ones
2. tenths
3. hundredths
4. tenths
5. ones
6. hundredths
7. tens
8. tenths
9. hundreds
10. tenths
11. hundredths
12. tens
13. thousandths
14. hundreds
15. tenths
16. thousandths
17. hundredths
18. tens

Page 17: Identify the Digit

1. 927.452
2. 469.36
3. 2,314.067
4. .94
5. 97.8
6. .056
7. 15.04
8. .458
9. 321.45
10. 532.497

Page 18: Zeros in Decimals

1. .230 .2300
2. .0460 .046
3. .10.100
4. 3.50 3.500
5. .070 .07
6. 9.080 9.08

Page 19: Writing Zeros to Hold Place Values

1. .037
2. .06
3. .9
4. .16
5. .427
6. .001
7. 1.014
8. 16.09
9. 427.3
10. 7.007
11. .05
12. .3

Page 20: Words to Decimals

	hundreds	tens	ones	and	tenths	hundredths	thousandths
1. Six and four tenths			6	.	4		
2. Fifteen and six hundredths		1	5	.	0	6	
3. Three hundred forty-five thousandths				.	3	4	5
4. One hundred and twenty-one hundredths	1	0	0	.	2	1	
5. Eight hundred ninety-four and six tenths	8	9	4	.	6		
6. Seven thousandths				.	0	0	7
7. Ninety-three and fifty-four thousandths		9	3	.	0	5	4
8. Seven and four tenths			7	.	4		
9. Two hundred seven and seventy-one hundredths	2	0	7	.	7	1	
10. Three thousandths				.	0	0	3
11. Five hundred thirty-nine thousandths				.	5	3	9
12. Nine hundred forty-five and three hundredths	9	4	5	.	0	3	
13. Three and eight thousandths			3	.	0	0	8
14. Eight tenths				.	8		
15. One hundred five and one hundredth	1	0	5	.	0	1	

Page 21: Find the Number

1. d
2. b
3. a
4. c
5. c
6. b
7. a
8. d
9. c
10. d
11. a
12. b
13. b
14. c
15. d
16. a
17. b
18. c
19. d
20. a

Page 22: Expanding Decimals

1. .73 = .7 + .03
2. .984 = .9 + .08 + .004
3. 8.275 = 8 + .2 + .07 + .005
4. .761 = .7 + .06 + .001
5. 75.94 = 70 + 5 + .9 + .04
6. 5.693 = 5 + .6 + .09 + .003
7. .94 = .9 + .04
8. 4.386 = 4 + .3 + .08 + .006

Page 23: Comparing Decimals

1. $>$
2. $<$
3. $=$
4. $>$
5. $<$
6. $<$

Page 24: Ordering Decimals

1. 4.2 4.25 4.279
2. 4.9 5.03 5.841
3. 7.01 7.032 7.4
4. 15.98 15.981 17.8
5. 94.3 94.301 94.31
6. 13.05 13.5 13.58
7. 3.05 3.50 5.30
8. .378 3.49 3.5

Page 25: Decimal Place-Value Review

1. greater
2. tens
3. tenths
4. thousandths
5. hundredths
6. .375, 3.75, 37.5
7. 93.2
8. 932
9. 9.032
10. $>$

Page 26: Steps for Addition

A. 6.1
B. 12.2
1. 7.7
2. 10.5
3. 5.7
4. 13.2
5. 15.1
6. 8.3
7. 11.3
8. 12.3
9. 11.0 or 11

Page 27: Adding Tenths, Hundredths, and Thousandths

A. $$\begin{array}{r} \scriptstyle(1) \\ 0.74 \\ +\ 0.32 \\ \hline 1.06 \end{array}$$

B. $$\begin{array}{r} \scriptstyle(1) \\ 0.845 \\ +\ 0.653 \\ \hline 1.498 \end{array}$$

1. $$\begin{array}{r} \scriptstyle(1) \\ 0.5 \\ +\ 0.6 \\ \hline 1.1 \end{array}$$

2. 1.0
3. 1.2
4. 1.61
5. .87
6. .11
7. 1.062
8. .502
9. 1.626

Page 28: Adding Mixed Decimals

A. $$\begin{array}{r} \scriptstyle(1)\,(1) \\ 9.67 \\ +\ 5.94 \\ \hline 15.61 \end{array}$$

B. $$\begin{array}{r} 7.03 \\ +\ 9.25 \\ \hline 16.28 \end{array}$$

1. $$\begin{array}{r} \scriptstyle(1)\,(1) \\ 3.97 \\ +\ 8.37 \\ \hline 12.34 \end{array}$$

2. 10.09
3. 12.00 or 12
4. 11.61
5. 8.55
6. 13.61
7. 6.37
8. 6.71
9. 16.53

Page 29: Think Zero

A. $$\begin{array}{r} \scriptstyle(1)(1) \\ 47.8 \\ +\ 2.3 \\ \hline 50.1 \end{array}$$

B. $$\begin{array}{r} \scriptstyle(1)(1) \\ 397.2 \\ +\ 85.4 \\ \hline 482.6 \end{array}$$

1. $$\begin{array}{r} \scriptstyle(1)(1) \\ 97.8 \\ +\ 3.2 \\ \hline 101.0 \end{array}$$

2. 17.1
3. 47.5
4. 67.0 or 67
5. 39.9
6. 49.9
7. 905.7
8. 543.6
9. 41.7

Page 30: Zeros as Placeholders

A. $$\begin{array}{r} \scriptstyle(1) \\ 5.48 \\ +\ 9.80 \\ \hline 15.28 \end{array}$$

B. $$\begin{array}{r} \scriptstyle(1)(1) \\ 34.60 \\ +\ 9.47 \\ \hline 44.07 \end{array}$$

C. $$\begin{array}{r} \scriptstyle(1) \\ 4.932 \\ +\ 1.400 \\ \hline 6.332 \end{array}$$

1. $$\begin{array}{r} \scriptstyle(1) \\ 7.45 \\ +\ 2.90 \\ \hline 10.35 \end{array}$$

2. 10.33
3. 44.16
4. 33.05
5. 71.88
6. 6.76
7. 13.074
8. 40.419
9. 10.583

Page 31: Writing Decimal Points

A. 61.
B. 522.
C. 144.
D. 2.
1. 11.38
2. 24.93
3. 33.9
4. 9.16
5. 22.75
6. 16.63

Page 32: Write Zeros When Necessary

A. 80.8
B. 386.32
1. 9.9
2. 19.99
3. 27.401
4. 17.94
5. 7.56
6. 32.93
7. 14.91
8. 19.7
9. 16.10

Page 33: Lining Up Decimals

1. $$\begin{array}{r} 613.049 \\ .620 \\ .953 \\ +\ \ \ .900 \\ \hline \end{array}$$

2. $$\begin{array}{r} 2.950 \\ 62.000 \\ 45.390 \\ +\ \ 3.846 \\ \hline \end{array}$$

3. $$\begin{array}{r} 312.749 \\ 51.090 \\ +\ \ 2.600 \\ \hline \end{array}$$

4. $$\begin{array}{r} \$21.40 \\ \$5.07 \\ +\ \$51.00 \\ \hline \end{array}$$

5. $$\begin{array}{r} 5.460 \\ 93.700 \\ 2.467 \\ +\ 45.060 \\ \hline \end{array}$$

6. $$\begin{array}{r} 3.200 \\ .985 \\ .460 \\ +\ \ 9.000 \\ \hline \end{array}$$

Page 34: Practice Helps

1. 28.543
2. 18.87
3. 15.526
4. 6.60 or 6.6
5. 6.86
6. 82.53
7. 9.849
8. 121.87

Page 35: Addition Review

1. 2.040
2. 18.032
3. <
4. 7.05, 7.105, 7.15
5. .549
6. 1.3
7. 6.228
8. 2.77
9. 62.357
10. 10.334
11. 3.231
12. 17.046

Page 36: Decimals and Money

1. c
2. b
3. f
4. d
5. a
6. e

Page 37: Working with Money

1. $5.00 + .25 + .05 + .01 + .01 = $5.32
2. $.05 + .10 + .10 + .05 + .25 = $.55
3. $.25 + .25 + .25 + .10 + .05 = $.90
4. $10.00 + .10 + .10 + .10 + .01 = $10.31

Page 38: Writing Number Sentences

1. $7.45 + $3.15 = $10.60
2. $13.05 + $5.62 = $18.67
3. $75.11 + $9.28 = $84.39
4. $45.00 + $7.99 = $52.99
5. $.68 + $6.45 = $7.13
6. $9.98 + $15.00 = $24.98

Page 39: Number Sentences

1. $17.35 + $6.92 = $24.27
2. $15.92 + $1.32 = $17.24
3. .45 + .39 = .84 pound
4. $25.50 + $19.75 = $45.25
5. 3.4 + .6 = 4.0 miles
6. $14.35 + $9.35 = $23.70

Page 40: Write a Question

Answers will vary but should resemble the following:

1. How much money did they spend in all?
2. How much did Shawna and her sister save altogether?
3. How much did Ernesto spend in all?
4. How much were all 3 CDs?
5. How much were both telephone bills together?
6. a) How much did they spend together?
b) How much did Sal and José spend?

Page 41: Using Symbols

1. a) 439.99 > 349.99
b) 349.99 < 439.99
c) 349.99 ≠ 439.99
2. a) 35.59 + 44.54 = 80.13
b) 35.59 < 44.54
c) 44.54 > 35.59
d) 35.59 ≠ 44.54

Page 42: Two-Step Word Problems

A. $379 + $1.25 = $5.04
B. $5.04 + $1.00 = $6.04
1. a) 171
b) 163
c) Gina
2. a) 18.9
b) 22.9
c) Lorena
d) 22.9 > 18.9

Page 43: Addition Word Problems

1. $35.99 + $29.00 = $63.99
2. $129.99 + $45.00 = $174.99
3. $14.35 + $2.80 = $17.15
4. 49.5 + 55.6 = 105.1
5. $445.80 + $129.89 = $575.69
6. $59.39 + $29.85 = $89.24
7. $195.77
8. a) $101.75
b) yes

Page 44: Addition Problem-Solving Review

1. $11.83
2. $6.07
3. $106.95
4. 34,007.11 miles
5. $11.59

Page 45: Steps for Subtraction

A. 4.2
B. .34
1. .2
2. .4
3. .6
4. 3.3
5. 1.3
6. 5.7
7. .11
8. .23
9. .21

Page 46: Regrouping

A. .07
B. 1.8
C. 3.92
1. .08
2. .38
3. .23
4. 5.8
5. 4.8
6. 1.7
7. 7.75
8. 1.56
9. 4.82

Page 47: Mastering Regrouping

A. .609
B. .578
C. 3.36
1. .288
2. .358
3. .029
4. .399
5. .157
6. .267
7. 5.49
8. 3.17
9. 1.39

Page 48: More Regrouping

A. 3.39
B. 5.28
C. 1.97
1. 4.34
2. 8.57
3. 6.69
4. 7.15
5. 5.02
6. 3.72
7. 1.54
8. 2.67
9. 6.79

Page 49: Zeros Help

1. .484
2. .463
3. 4.345
4. .062
5. 7.093
6. .261
7. .296
8. 7.062
9. .417
10. .195
11. .492
12. 4.637

Page 50: Zeros at Work

A. 85.28
B. 64.27
C. 34.47
1. 46.86
2. 72.49
3. 40.66
4. 46.98
5. 51.38
6. 22.75
7. 50.15
8. 2.27
9. 63.65

Page 51: Subtracting Decimals from Whole Numbers

A. 6.5
B. 1.18
C. 6.937
1. 1.9
2. 6.5
3. 14.3
4. 2.57
5. 17.94
6. 14.18
7. 2.982
8. 8.287
9. 4.976

Page 52: Subtracting Whole Numbers from Mixed Decimals

A. 25.5
B. 29.43
C. 52.849
1. 9.8
2. 46.1
3. 37.3
4. 37.19
5. 165.70
6. 8.62
7. 7.691
8. 18.512
9. 57.822

Page 53: Watch Out for Whole Numbers

A. 5.86
B. 67.2
C.

```
  (2)(14)(15)
    35.5
  -  8.9
  ------
    26.6
```

1. 7.24
2. 63.8
3. 4.22
4. 24.65
5. 13.7
6. 69.5
7. 9.70 or 9.7
8. 93.4
9. 6.08

Page 54: Add or Subtract?

1. 9.4
2. 51.9
3. 117.47
4. 38.86
5. 8.53
6. 11.74
7. 46.13
8. 20.94
9. 27.17
10. 54.9
11. 1.01
12. 710.67

Page 55: Putting It All Together

1. $14.5 < 14.6$
2. $7.82 > 7.6$
3. $32.73 < 33.2$
4. $67.4 > 66.73$
5. $47.9 > 47.605$
6. $30.4 = 30.4$
7. $142.51 < 144.88$
8. $1.29 = 1.29$

Page 56: Subtraction Review

1. .24
2. 2.4
3. .08
4. 1.19
5. .059
6. .41
7. 1.15
8. .558
9. .53
10. 26.275
11. 29.34
12. .633

Page 57: Number Sentences

1. 2.4 – .21 = 2.19 ounces
2. 118.6 + 69.8 = 188.4 miles
3. $32.45 – $9.50 = $22.95
4. $38.90 – $4.50 = $34.40
5. 15.8 – 2.9 = 12.9 gallons
6. 1.53 + .75 = 2.28 pounds
7. $75.95 – $22.75 = $53.20
8. $125.00 – $95.00 = $30.00

Page 58: Does the Answer Make Sense?

1. 101.4° – 98.6° = 2.8°; 2.8
2. 172.8 + 39.6 = 212.4; 212.4
3. $99.95 – $25.00 = $74.95; 74.95
4. $20.00 – $14.35 = $5.65; 5.65
5. $1,258.15 – $1,217.92 = $40.23; 40.23
6. 31.4 + 44.3 = 75.7; 75.7

Page 59: Using Symbols

1. a) $9.40 + $5.49 = $14.89
   b) $14.89 – $5.49 = $9.40
   c) $14.89 > $9.40
   d) $9.40 < $14.89
   e) $14.89 ≠ $9.40
2. a) $17.19 + $28.14 = $45.33
   b) $45.33 – $28.14 = $17.19
   c) $45.33 > $17.19
   d) $17.19 < $45.33
   e) $45.33 ≠ $17.19

Page 60: Write a Question

Questions should be similar to these:

1. a) How much did Leslie and Dawn spend altogether?
   b) How much more did Leslie spend than Dawn?
2. a) Sean saved how much more than his sister?
   b) How much did Sean and his sister save in all?
3. a) What was the total cost of the groceries and gasoline?
   b) The groceries cost how much more than the gasoline?
4. a) The jacket cost how much more than the sweater?
   b) What was the combined cost of the jacket and sweater?
5. a) Mr. Alatalo and Mr. Forbes saved how much altogether?
   b) Mr. Alatalo saved how much more than Mr. Forbes?
6. a) Mr. Fenwick paid how much in taxes both years combined?
   b) How much more did Mr. Fenwick pay in taxes this year than last?

Page 61: Decide to Add or Subtract

Answers will vary, but should represent a knowledge of addition and subtraction.

Page 62: Mixed Addition and Subtraction

1. $165.34 + $34.95 = $200.29; 200.29
2. $5.00 – $2.91 = $2.09; 2.09
3. 256.3 + 136.9 = 393.2; 393.2
4. 5.4 + 3.2 = 8.6; 8.6
5. $45.32 – $36.50 = $8.82; 8.82
6. $10.00 – $7.25 = $2.75; 2.75
7. $29.15 + $18.11 = $47.26; 47.26
8. $20.00 – $12.25 = $7.75; 7.75

Page 63: Two–Step Word Problems

A. $4.14

B. $15.86

1. a) $96.94 ⊞
b) $3.06 ⊟

2. a) $16.00 ⊟
b) $18.50 ⊞

3. a) 12.25 miles ⊞
b) 2.75 miles ⊟

4. a) $61.50 ⊞
b) $64.00 ⊞

5. a) $47.98 ⊞
b) $ 2.02 ⊟

6. a) $41.61 ⊞
b) $154.01 ⊟

Page 64: Two-Step Problem Solving

Questions will vary. Final answers are listed below.

1. $3.55
2. $111.93
3. $754.00
4. $43.76

Page 65: Multistep Word Problems

Questions will vary. Final answers are listed below.

1. Muriel received $1.89 in change.
2. The family had to pay $99.61.
3. Chris has 5.6 miles left to go.
4. Millie had $136.91 left.

Page 66: Picture Problems

1. $5.25
2. $18.51
3. $16.50
4. $1.63 + $4.33 + $2.92 = $8.88 total change
5. $2.60
6. $96.75

Page 67: Subtraction Problem-Solving Review

1. 7.88 lb
2. 1.2 lb
3. 4.1°
4. $10.75
5. $2.67
6. $19.75
7. $781.22
8. $96.74

Page 68: Reading Temperatures

a. 100.5°
b. 100.1°
c. 100.9°
d. 100.7°

1. 98.6°
2. 100.1°
3. 2.2°
4. 3.7°

Page 69: Figuring Change

A. $20.00 − $15.50 = $4.50

1. $10.00 − $6.20 = $3.80

2. $25.00 − $23.49 = $1.51

3. $50.00 − $43.50 = $6.50

4. $5.00 − $3.50 = $1.50

Page 70: Making Change

		$5	$1	25¢	10¢	5¢	1¢
			4			1	
1. $7.08	$20.00 − $12.92 = $7.08	1	2			1	3
2. $1.52	$5.00 − $3.48 = $1.52		1	2			2
3. $.17					1	1	2
4. $4.16			4		1	1	1
5. $7.59		1	2	2		1	4

Page 71: Money Problems

1. $.92
2. $251.41
3. $46.35
4. a) jacket and clock
   b) $.91 less
5. a) Sandwich
   b) Cola
   c) French Fries

Page 72: Keeping Records

1. 15.13 pounds
2. 6.01, 3.50, 3.23, 2.39
3. 3.62 pounds
4. 3.89 pounds
5. $58.94
6. a) Meals   b) Bait   c) Lodging

Page 73: Breaking the Record

1. 52.04
2. .44
3. a) Nancy 12.4
   b) Maria 12.58
   c) Shara 13.2
   d) Carrie 13.86
4. 1.46
5. Yes, .12 second

Page 74: Life-Skills Math Review

1. 3.8°
2. $1.50
3. $7.57; 1 5-dollar, 2 1-dollar, 2 quarters, 2 pennies
4. $106.67
5. 22 miles

Page 75: Cumulative Review

1. a) =
   b) >
2. 10.06, 10.6, 10.64
3. 19.723
4. 16.576
5. $23.83
6. .11
7. .078
8. .543
9. 1.33 < 1.69
10. $14.57

Page 76: Posttest

1. 66.01
2. $16.57
3. 1.49
4. 8
5. 25.16
6. $218.98
7. 51.85
8. 12.017
9. $2.62
10. 3.54
11. 11.307
12. $111.14
13. 5.15
14. .95 meter
15. 7.300
16. 8.25 pounds
17. 5.09, 5.2, 5.21
18. 12.1 miles
19. 83.29
20. 2.3 inches

Posttest Evaluation Chart

If a student misses one or more problems in a skill area, assign a review of the practice pages for that skill.

Skill Area	Posttest Problem #	Skill Section	Review Page
Place Value	4, 8, 15, 17	7–24	25
Addition	1, 5, 11, 19	26–34	35
Subtraction	3, 7, 10, 13	45–55	56
Addition Word Problems	6, 12, 16, 18	36–-43	44
Subtraction Word Problems	2, 9, 14, 20	57–66	67
Life-Skills Math	All	68–73	74

DECIMALS
Multiplication & Division

The approach for teaching multiplication and division of decimals is similar to that used for addition and subtraction of decimals. Teachers should develop an understanding of the concepts by proceeding from the concrete to the abstract and symbolic.

Student Glossary

Acquainting students with definitions of key math terms and life-skills concepts will enhance their mastery of the materials. Below are words defined in the student text. A glossary is provided at the end of the student text and on page 243 of the *Teacher's Resource Guide and Answer Key*.

average
brand
decimal place
dividend
divisor
down payment
mileage
monthly payment
number relation symbol
operational symbol
overtime
per
placeholder
quotient
regroup (carry)
remaining balance
temperature
total earnings
unit price
units of measure
whole number

STUDENT PAGE
10
Multiplication

Model Using a Number Line

A number line is a good model to illustrate the concept of multiplying decimals.

EXAMPLE If .3 + .3 = .6, then .3 × 2 = .6

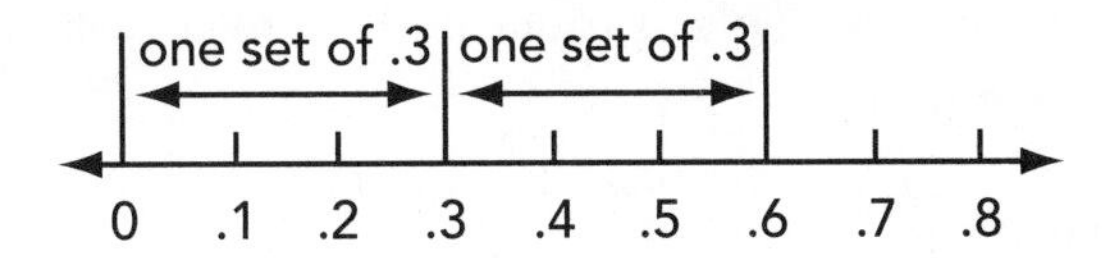

STUDENT PAGE
10
Multiplication

Use Drawings and Manipulatives

You can use drawings like the one below or manipulatives, such as clothespins or paper clips, to demonstrate multiplication of decimals.

EXAMPLE .3 + .3 = .6 so .3 × 2 = .6

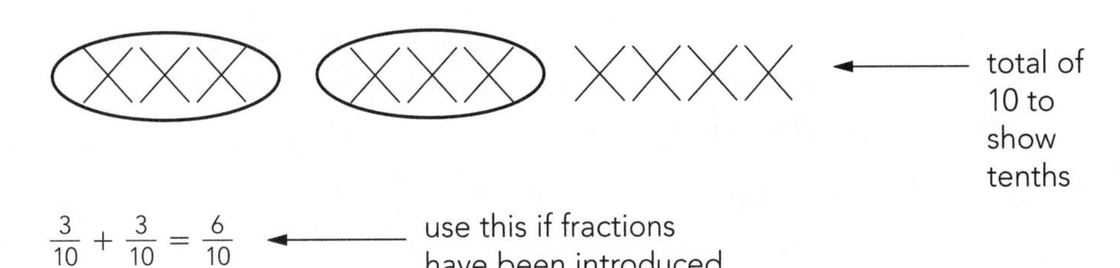

$\frac{3}{10} + \frac{3}{10} = \frac{6}{10}$ ← use this if fractions have been introduced

STUDENT PAGE
12
Multiplication

Draw a Diagram

Using a 10 × 10 grid or section of graph paper, let students diagram multiplication and division problems. After practicing with several drawings, show students that when decimals are multiplied, there are as many decimal places in the product as in both factors together.

EXAMPLE .6 × .7 = .42

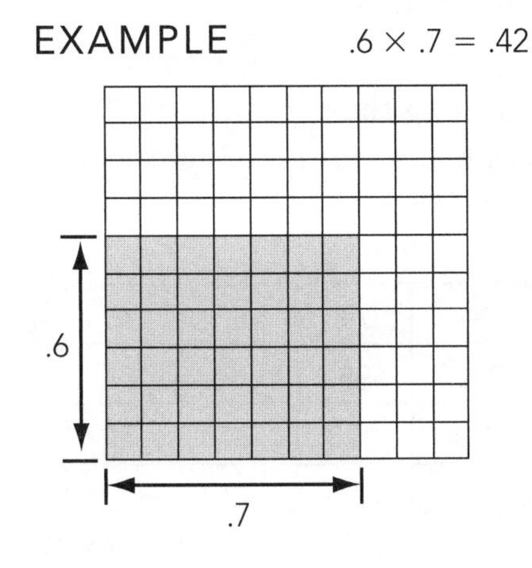

STUDENT PAGE
12
Multiplication

Relating Decimal and Fraction Multiplication

If students have already learned to multiply fractions, they can use this to learn about multiplying decimals.

1 Where both factors are written as decimals, have students rewrite the decimals as fractions. Tell them to multiply as usual, and the result will be the product expressed in fraction form.

EXAMPLES

$$\begin{array}{r} .4 \\ \times \quad .3 \\ \hline .12 \end{array} \quad \text{is the same as } \frac{4}{10} \times \frac{3}{10} = \frac{12}{100} = .12$$

$$\begin{array}{r} .06 \\ \times \quad .4 \\ \hline .024 \end{array} \quad \text{is the same as } \frac{6}{100} \times \frac{4}{10} = \frac{24}{1{,}000} = .024$$

2 After practicing with several examples, students should discover that when decimals are multiplied there are as many decimal places in the product as in both factors together.

3 When students understand this concept thoroughly and discover that the location of the decimal point can be determined by counting the number of places in the factors, they are ready to follow the rule that is based on counting decimal places.

NOTE: Point out that decimal points do not have to be aligned in multiplication since the factors are to be multiplied as if they were whole numbers and the product is then changed to a decimal by placing a decimal point in the correct position in the answer.

STUDENT PAGE
13
Multiplication

Locate the Decimal Point

To review the placement of the decimal point in the product, use a grid. You may place more numbers along the top and side. Tell students that they must count the decimal places being multiplied on the outside of the grid and place decimal points on the numbers inside.

EXAMPLE

X	.35	7.6	8	10
9.2	3220	6992	736	920
.3	105	228	24	30

STUDENT PAGE
15
Multiplication

Mental Multiplication

Using flash cards, let students respond orally or in writing. The computations are simple enough that students can calculate and fix the decimal point mentally.

3 × .03 = _______	.007 × .8 = _______
.06 × .08 = _______	.06 × .8 = _______
.7 × .6 = _______	.9 × .6 = _______
.009 × 2 = _______	4 × .03 = _______

STUDENT PAGE
16
Multiplication

Estimate the Answer

1 Have students estimate the product of a whole number and a decimal. Then have them work the exact calculation and compare results.

EXAMPLE

$$\begin{array}{r} 6.45 \\ \times\ 8 \\ \hline 51.60 \end{array} \quad \text{Round 6.45 to 6} \quad \begin{array}{r} 6 \\ \times\ 8 \\ \hline 48 \end{array}$$

Practice with many examples. This will provide quick answers when dealing with estimation and problem solving. Use money amounts.

EXAMPLE $9.45 × 8 = _______
9 × 8 = 72

2 Let students make up story problems using words like *almost, approximately,* and *estimate* to get a sense of story problems that need estimated answers.

Alternate Activities

1 $8.24 × $5.25 = _______
8 × 5 = _______

Work out exact answers after students get rounded answers. Compare results.

2 Circle the best estimated answer for each problem.

Problem	Answer Choices		
9.32 × 5.2 =	45	450	4,500
2.2 × 4.1 =	8	80	800
7.3 × 2.695 =	21	210	2,100

STUDENT PAGE
21
Multiplication

Multiplying by 10, 100, and 1,000

Have students replace the □ with 10, 100, or 1,000 to make each number sentence true.

EXAMPLES

4.5 × □ = 45

.065 × □ =65

9.32 × □ = 9,320

34.8 × □ = 348

STUDENT PAGE
24
Multiplication
Problem
Solving

Substitute Different Numbers

Let students generate different numbers for the same problem.

EXAMPLES

House paint costs $8.78 per gallon.
How much will 4.5 gallons cost?

House paint costs $10.50 per gallon.
How much will 3 gallons cost?

House paint costs $______ per gallon.
How much will ______ gallons cost?

STUDENT PAGE
25
Multiplication
Problem
Solving

High-Interest Problem Solving

Let your students become active participants in creating and solving real-life problems.

Have students:

1. Fill in the blanks with reasonable numbers.
2. Write a question about the facts.
3. Complete a number sentence for each problem.
4. Write the answer in sentence form. This helps students clarify their thinking—"Does the answer make sense?"

EXAMPLE

Ethan mailed ________ packages that weighed ________ pounds each.

Question: ____

______ ______ ______ = ______
Number Operation Symbol Number Answer

Answer: __________

STUDENT PAGE
40
Division

Mental Division

Using flash cards, let students ask and answer division questions. They should respond orally. Make sure the computations are simple enough for the students to calculate quickly.

EXAMPLES

$3\overline{).06}$	$7\overline{)5.6}$	$.5\overline{)2.5}$
$.09\overline{).063}$	$.08\overline{).008}$	$.6\overline{)5.4}$

STUDENT PAGE
45
Division

Rounding Money Amounts

Often when we divide money amounts, we have more digits in the answer than we need. Students can practice rounding money amounts to the nearest cent.

Step 1: Circle the cents.
$15.(83)6

Step 2: If the digit that follows the circled amount is 5 or more, add one cent to the circled amount. If the digit is 4 or less, the circled amount stays the same.

Step 3: Write the rounded amount. $15.84

EXAMPLES

$4.935 rounds to $4.94.

$25.852 rounds to $25.85.

$2.698 rounds to ________.

$20.338 rounds to ________.

STUDENT PAGE
46
Division

Dividing by 10, 100, and 1,000

Have students replace the □ with 10, 100, or 1,000 to make each statement true.

EXAMPLES

9.05 ÷ □ = .905

3 ÷ □ = .3

1.5 ÷ □ = .0015

80 ÷ □ = 8

STUDENT PAGE
46
Division

Multiplying and Dividing by Power of Ten

Have students complete the pattern for each of the sequences.

EXAMPLES

7, .7, .07, ________, ________
What is the pattern?________

Answer: Each number is divided by 10.

.8, 8, 80, ________, ________
What is the pattern?________

8, .8, .08, ________, ________
What is the pattern?________

STUDENT PAGE
49
Division

Decimals at Work

Review the four operations with decimals. Have students use the decimal numbers to find the sum, difference, product, and quotient.

EXAMPLES

A 3.6 and .9
Sum ________
Difference ________
Product ________
Quotient ________

B 14.56 and 1.3
Sum ________
Difference ________
Product ________
Quotient________

STUDENT PAGE
49
Division

Relating Operations

Have students complete the number sentences to make each statement true. First model how you would use trial and error to figure out what to do.

EXAMPLES

		Operation Symbol	Decimal	=		
A	6.3	÷	2.1	=	3	6.3 ÷ 3 = 2.1
B	3.04	x	5	=	15.2	15.2 ÷ 5 = 3.04
C	.045 (Decimal)	+ (Operation Symbol)	.05	=	.095	.095 − .05 = .045
D	18.32 (Decimal)	− (Operation Symbol)	2.98	=	15.34	15.34 + 2.98 = 18.32

STUDENT PAGE
51
Division
Problem
Solving

Everyday Problems

Write a real-life problem on the board or overhead projector that your students can relate to.

EXAMPLE Sam earned $107.25 at $8.25 per hour. How many hours did he work?

After students solve the problem, substitute different amounts. Discuss how to round the answer to the nearest cent.

Problems with money and pay provide more interesting practice than simply assigning a set of division problems from a textbook.

STUDENT PAGE
52
Division
Problem
Solving

Find the Product and Quotient

Using the two numbers in the boxes, write number sentences to find the product and quotient.

EXAMPLE

[7.2] [1.2] Answers: 7.2 ÷ 1.2 = ______

7.2 × 1.2 = ______

STUDENT PAGE

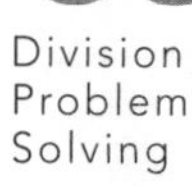

Division Problem Solving

Generate Your Own Decimal Story Problems

Let your students become active participants in creating and solving real-life problems.

Have students:

1. Fill in the blanks with reasonable decimal numbers.
2. Write a question about the facts.
3. Complete a number sentence for each problem.
4. Write the answer in sentence form. This helps students clarify their thinking—"Does the answer make sense?"

EXAMPLE Kelly drove ______ miles on ______ gallons of gasoline. How many miles did she get per gallon?

______ ______ ______ = ______

Number | Operation Symbol | Number | Answer

Answer: ____________

STUDENT PAGE

59

Division Problem Solving

Locating Information to Answer Questions

Using fliers from major department stores, students can follow written or oral instructions to locate information or answer questions.

EXAMPLES

Turn to any page in the flier. Ask questions such as:

What is the cost of the most expensive item on the page?

What is the cost of the least expensive item on the page?

About how many times greater is the most expensive item than the least expensive item?

How much would the 5 most expensive items cost?

What is the difference between the least and the most expensive item on the page?

What is the difference in the regular price and the sale price of one of the items?

You can easily create questions that include problems using addition, subtractions, multiplication, and division of decimals. With this activity students gain real-life number sense and consumer awareness.

Page 4: Pretest

1. .0492
2. 3.33
3. 249.66
4. .1058
5. .0632
6. 3.2
7. 124
8. .04
9. 9
10. 60
11. 400
12. 3.2
13. .491
14. .086
15. $286.92
16. 3 pounds
17. 18 miles
18. 219.5 miles
19. 35 hours
20. 520 pounds

Pretest Evaluation Chart

If a student misses any problems in a skill area, assign the practice pages for that skill. However, students may need to complete all practice pages to reinforce areas of weakness.

Skill Area	Pretest Problem #	Skill Section	Review Page
Multiplication	1, 2, 3, 4, 5, 6, 7	7–22	23, 50
Division	8, 9, 10, 11, 12, 13, 14	32–48	49, 50
Multiplication Problem Solving	15, 17, 20	24–30	31
Division Problem Solving	16, 18, 19	51–64	65
Life-Skills Math	All	66–73	74

Page 7: Counting Decimal Places

1. 2
2. 1
3. 0
4. 2
5. 2
6. 1
7. 3
8. 0
9. 3
10. 1
11. 3
12. 1
13. 2
14. 2
15. 3
16. 0

Page 8: Decimal Places in Answers

1. 2 + 2 = 4
2. 2 + 1 = 3
3. 3 + 1 = 4
4. 0 + 2 = 2
5. 3 + 2 = 5
6. 2 + 0 = 2
7. 2 + 1 = 3
8. 1 + 2 = 3

Page 9: Place the Decimal Point

1. 83.4
2. 4.027
3. 1.6425
4. .97
5. 683
6. 9.06
7. .7513
8. 17
9. 2.374
10. 7.4

11.
```
  7.5   1
×  .3   1
 2.25   2
```

12.
```
 2.47   2
×  .4   1
 .988   3
```

13.
```
 .906   3
×  .7   1
.6342   4
```

14.
```
  .89   2
×  53   0
  267
 445
47.17   2
```

Page 10: Tenths Times a Whole Number

A. 0.9

B.
```
  0.3   1
×   3   0
  0.9   1
```

1. a) 1.2

b)
```
  0.4   1
×   3   0
  1.2   1
```

2. a) 18.4

b)
```
  4.6   1
×   4   0
 18.4   1
```

3. a) 113.8

b)
```
  56.9   1
×    2   0
 113.8   1
```

4. a) 973.5

b)
```
 324.5   1
×    3   0
 973.5   1
```

DECIMALS: Multiplication and Division

Page 11: Hundredths Times a Whole Number

A. .75

B.

0.25	2
× 3	0
0.75	2

1. a) 1.14

b)

0.38	2
× 3	0
1.14	2

2. a) 24.20

b)

6.05
× 4
24.20

3. a) 184.82
b) 184.82

4. a) 1,420.86
b) 1,420.86

Page 12: Multiplying Without Regrouping

1.	.9 1 × 7 1 .63 2	**4.**	6.21 2 × .4 1 2.484 3	**7.**	.534 3 × .2 1 .1068 4
2.	2.3 1 × .3 1 .69 2	**5.**	.403 3 × .3 1 .1209 4	**8.**	8.23 2 × .2 1 1.646 3
3.	.05 2 × 9 0 .45 2	**6.**	.8 1 × .4 1 .32 2	**9.**	11.2 1 × .4 1 4.48 2

Page 13: Multiplication with Regrouping

1. .120
2. 2.94
3. 39.9
4. 1.32
5. 4.98
6. 2.454
7. .987
8. .3661
9. .2152
10. 39.70
11. 1.704
12. 294.4

Page 14: Zeros as Placeholders

A.	.06 2 × .8 1 .048 3	**3.**	.009 3 × 4 0 .036 3
B.	.17 2 × .03 2 .0051 4	**4.**	.003 3 × .06 2 .00018 5
1.	.007 3 × .5 1 .0035 4	**5.**	.135 3 × .7 1 .0945 4
2.	.34 2 × .02 2 .0068 4	**6.**	.26 2 × .009 3 .00234 5

Page 15: Practice Adding Zeros

1.	.08 2 × .4 1 .032 3	**7.**	.027 3 × .05 2 .00135 5
2.	.006 3 × 7 0 .042 3	**8.**	.099 3 × .02 2 .00198 5
3.	.29 2 × .06 2 .0174 4	**9.**	.642 3 × .018 3 .011556 6
4.	.026 3 × .8 1 .0208 4	**10.**	1.14 2 × .03 2 .0342 4
5.	.302 3 × .03 2 .00906 5	**11.**	9.89 2 × .002 3 .01978 5
6.	.54 2 × .17 2 .0918 4	**12.**	.25 2 × .35 2 .0875 4

Page 16: Line Up and Multiply

1. $\begin{array}{r} 1.4 \\ \times\ 8 \\ \hline 11.2 \end{array}$ **2.** $\begin{array}{r} 6.7 \\ \times\ 5 \\ \hline 33.5 \end{array}$ **3.** $\begin{array}{r} 2.6 \\ \times\ 3 \\ \hline 7.8 \end{array}$

4. $\begin{array}{r} 2.3 \\ \times\ .4 \\ \hline .92 \end{array}$ **5.** $\begin{array}{r} 4.7 \\ \times\ 5 \\ \hline 23.5 \end{array}$ **6.** $\begin{array}{r} 8.9 \\ \times\ 6 \\ \hline 5.34 \end{array}$

7. $\begin{array}{r} .84 \\ \times\ .8 \\ \hline .672 \end{array}$ **8.** $\begin{array}{r} .12 \\ \times\ .7 \\ \hline .084 \end{array}$ **9.** $\begin{array}{r} .45 \\ \times\ .2 \\ \hline .090 \end{array}$

Page 17: Longer Number on Top

A. $\begin{array}{r} 1.3 \\ \times\ .7 \\ \hline .91 \end{array}$ **B.** $\begin{array}{r} 7.9 \\ \times\ .5 \\ \hline 3.95 \end{array}$

1. $\begin{array}{r} 6.2 \\ \times\ 8 \\ \hline 49.6 \end{array}$ **2.** $\begin{array}{r} 25 \\ \times\ .7 \\ \hline 17.5 \end{array}$ **3.** $\begin{array}{r} 4.9 \\ \times\ 3 \\ \hline 14.7 \end{array}$

4. $\begin{array}{r} 16 \\ \times\ .4 \\ \hline 6.4 \end{array}$ **5.** $\begin{array}{r} 2.7 \\ \times\ .6 \\ \hline 1.62 \end{array}$ **6.** $\begin{array}{r} 8.6 \\ \times\ .1 \\ \hline .86 \end{array}$ **7.** $\begin{array}{r} 1.8 \\ \times\ .9 \\ \hline 1.62 \end{array}$ **8.** $\begin{array}{r} 17 \\ \times\ .4 \\ \hline 6.8 \end{array}$

9. $\begin{array}{r} 3.3 \\ \times\ .5 \\ \hline 1.65 \end{array}$ **10.** $\begin{array}{r} 4.8 \\ \times\ 6 \\ \hline 28.8 \end{array}$ **11.** $\begin{array}{r} 7.9 \\ \times\ .3 \\ \hline 2.37 \end{array}$ **12.** $\begin{array}{r} 6.9 \\ \times\ .2 \\ \hline 1.38 \end{array}$

Page 18: Multiplying by Hundredths

A. $\begin{array}{r} .56 \\ \times\ .63 \\ \hline 168 \\ 3360 \\ \hline .3528 \end{array}$ **B.** $\begin{array}{r} .45 \\ \times\ .14 \\ \hline 180 \\ 450 \\ \hline .0630 \end{array}$ **C.** $\begin{array}{r} 7.2 \\ \times\ .38 \\ \hline 576 \\ 2160 \\ \hline 2.736 \end{array}$

1. .5655
2. 279.3
3. 15.12
4. .0925
5. .817
6. 34.08
7. 9.114
8. 1.608
9. 56.42

Page 19: Practice Helps

1. $\begin{array}{r} 4.35 \\ \times\ .38 \\ \hline 3480 \\ 13050 \\ \hline 1.6530 \end{array}$
2. 5.396
3. 18.565
4. .40426
5. $\begin{array}{r} .701 \\ \times\ .43 \\ \hline 2103 \\ 28040 \\ \hline .30143 \end{array}$
6. .02888
7. 406.00
8. .03151
9. $\begin{array}{r} 9.45 \\ \times\ 5.6 \\ \hline 5670 \\ 47250 \\ \hline 52.920 \end{array}$
10. 382.28
11. 1.6544
12. 140.55

Page 20: Put It All Together

1. 2.44
2. 5.7
3. .738
4. 1.050
5. 256.23
6. 1.242
7. 33.75
8. .00584
9. .04515
10. 330.33
11. .054
12. 16.89

Page 21: Multiplying by 10, 100, and 1,000

1. 59.6
2. 6. or 6
3. .4
4. 23.84
5. 8.03
6. 326. or 326
7. 68. or 68
8. 435. or 435
9. 12.3
10. 602.7
11. 42. or 42
12. 6,019. or 6,019
13. 1,093. or 1,093
14. 721. or 721
15. 3,224. or 3,224

Page 22: Practice Your Skills

1. 2,400
2. 10
3. 8,270
4. 914.5
5. 425
6. .7
7. 90
8. 34,720
9. 88,120
10. 702.35
11. 7.07
12. 860
13. 33.0
14. 6,260
15. 101.0

Page 23: Multiplication Review

1. 2

2.

.06	2
× 8.4	1
	3

3.

5.43	2
× .7	1
3.801	3

4. a) 156.3

b)

52.1	1
× 3	0
156.3	1

5. .0912
6. 67.28

7.

.002	3
× .09	2
.00018	5

8.

.16	2
× .04	2
.0064	4

9. 1.4
10. 34.4
11. .432
12. 2.116
13. 19.44
14. 1.6544
15. 864.3
16. 63.4

Page 24: Number Sentences

1. $3.35 × 5 = $16.75; 16.75
2. $3.45 × 7 = $24.15; 24.15
3. 4.8 × $.89 = $4.56; 4.56
4. $18.78 × 4.5 = $84.51; 84.51
5. $195.64 × 3 = $586.92; 586.92
6. $15.75 × 5 = $78.15; 78.15

Page 25: Does the Answer Make Sense?

1. 45.5 × 5.5 = 250.25; 250.25
2. 4 × $1.25 = $5.00; 5.00
3. 9.6 × $12.45 = $119.52; 119.52
4. $10.75 × 52 = $559.00; 559.00
5. 3.4 × 6 = 20.4; 20.4
6. 2.5 × 3 =7.5; 7.5

Page 26: Grocery Shopping

1. a) $4.38
b) $4.05
c) $2.37

2. a) $5.18
b) $3.98
c) $4.58

3. $24.54

Page 27: Think It Through

1. a) How much will 4 slices of pizza cost? $1.25 × 4 = $5.00
b) How much will 6 slices of pizza cost? $1.25 × 6 = $7.50

2. a) How many cars are in the garage? 8 × 4 = 32
b) How many cars are on 2 floors? 8 × 2 = 16

3. a) How much are 2 movie tickets? $8.50 × 2 = $17.00
b) How much are 4 movie tickets? $8.50 × 4 = $34.00

4. a) How many people went to the game? 4 × 8 = 32
b) How much money was raised to pay for the vans? 8 × $10.00 = $80.00

5. a) How many students helped to raise money? 20 × 10 = 200
b) How much money was raised by each class? 20 × $50.00 = $1,000.00

6. a) How much was made from selling $5.50 tickets? 25 × $5.50 = $137.50
b) How much was made from selling $5.00 tickets? 40 × $5.00 = $200.00

Page 28: Write a Question

1. a) How much did all the mushrooms cost?
   b) $4 \times \$3.29 = \$13.16$
2. a) How much did all the soda cost?
   b) $\$.75 \times 15 = \$11.25$
3. a) How much did Jane earn?
   b) $\$8.80 \times 19 = \$162.70$
4. a) How far can the car drive?
   b) $17.8 \times 15 = 267$ miles
5. a) How far did the airplane fly?
   b) $523.5 \times 5 = 2{,}617.5$ miles
6. a) How far did Avishan run?
   b) $4 \times 2 = 8$ miles

Page 29: Using Symbols

1. $20.25 = 20.25$
2. $6.0 < 6.2$
3. $43.79 > 42.16$
4. $16.4 > 16$
5. $23.4 < 25.00$
6. $18.34 < 19.866$
7. $12.65 > 14.75$
8. $101 > 100.1$

Page 30: Apply Your Skills

1. $\$4.57 \times 5 = \$22.85$; 22.85
2. $\$44.59 + \$22.39 = \$66.98$; 66.98
3. $\$259.67 - \$39.87 = \$255.80$; 255.80
4. $\$886.57 + \$669.79 = \$1{,}556.36$; 1,556.36
5. $\$3.99 \times 9 = \$35.91$; 35.91
6. $\$27.54 - 23.43 = \$4.11$; 4.11

Page 31: Multiplication Problem-Solving Review

1. $225.00
2. $\$35.85 \times 12 = \$430.20$; 430.20
3. $\$11.22 \times 25 = \$280.50$; 280.50
4. $10.95
5. 248 pieces of candy
6. 448 miles
7. $300.00
8. $61.20

Page 32: Understanding Division Words

A. quotient
B. dividend
C. divisor
1. 2
2. 8
3. 16
4. 24
5. 4
6. 6
7. 36
8. 4
9. 9
10. 8
11. 48
12. 6
13. 63
14. 7
15. 9
16. 35
17. 5
18. 7

Page 33: Dividing a Decimal by a Whole Number

1. $.25

$$\begin{array}{r} \$.25 \\ 2\overline{)\$.50} \\ \underline{4} \\ 10 \\ \underline{10} \\ 0 \end{array}$$

2. $.25

$$\begin{array}{r} \$.25 \\ 3\overline{)\$.75} \\ \underline{6} \\ 15 \\ \underline{15} \\ 0 \end{array}$$

3. $.10

$$\begin{array}{r} \$.10 \\ 4\overline{)\$.40} \\ \underline{4} \\ 00 \\ \underline{0} \\ 0 \end{array}$$

4. $.16

$$\begin{array}{r} \$.16 \\ 5\overline{)\$.80} \\ \underline{5} \\ 30 \\ \underline{30} \\ 0 \end{array}$$

Page 34: Move the Decimal Straight Up

1. .3
2. .3
3. 1.5
4. .308
5.
```
    .46
 7)3.22
   28
    42
    42
     0
```
6. .5
7. .108
8. .6
9.
```
    8.7
 4)34.8
   32
    28
    28
     0
```
10. .47
11. .67
12. .136

Page 35: Zeros as Placeholders

1. .02
2. .005
3. .06
4. .09
5. .001
6.
```
   .007
 8).056
     0
     56
     56
      0
```
7. .07
8. .004
9. .002
10. .07
11.
```
   .03
 7).21
     0
    21
    21
     0
```
12. .09
13. .004
14. .03
15. .08

Page 36: Dividing by a Decimal

No student problems on this page.

Page 37: Place the Decimals

```
1.          14.6
      .3 )4.3↑8
     ×10     ×10

2.           .65
     .96 ).62↑40
    ×100    ×100

3.           .32
     4.9 )1.5↑68
     ×10     ×10

4.            .6
    .175 ).105↑0
   ×1000   ×1000

5.           1.8
      .9 )1.6↑2
     ×10     ×10

6.           5.7
     .58 )3.30↑6
    ×100    ×100

7.          45.6
     .04 )1.82↑4
    ×100    ×100

8.          78.9
    .007 ).552↑3
   ×1000   ×1000

9.           2.5
     1.9 )4.7↑5
     ×10     ×10
```

Page 38: Dividing by Tenths

1.
```
      .16
 3.3).528
      33
      198
      198
        0
```
2. 6
3. .04
4. 8
5. .92
6. 3.4
7. .82
8. 4.5
9. .27

Page 39: Dividing by Hundredths

1.
```
      63.
 .07)4.41
     42
      21
      21
       0
```
2. 4.1
3. 3
4. 4.1
5. 40.9
6. 14
7. 4.5
8. 36
9. 292.1

Page 40: Dividing by Thousandths

1. 256
2. 7.8
3. 4,701
4. 1,406
5. 139
6. 789
7. 23.1
8. 89
9. 14.7

Page 41: Mixed Practice

1. 7	**5.** 10.6	**9.** .13
2. 70.9	**6.** .08	**10.** 285.2
3. 32	**7.** 2.4	**11.** .47
4. 14.7	**8.** .8	**12.** 40

Page 42: Zeros in the Dividend

1. 50	**4.** 428	**7.** 85
2. 85	**5.** 75	**8.** 20
3. 245	**6.** 800	**9.** 2,635

Page 43: Whole Numbers Divided by Decimals

1. 5	**5.** 325	**9.** 125
2. 30	**6.** 225	**10.** 625
3. 75	**7.** 700	**11.** 750
4. 2	**8.** 1,200	**12.** 12,000

Page 44: Zeros in the Quotient

A.
```
    .02
.8).016
    16
     0
```

B.
```
        .009
56.5).5085
      5085
         0
```

C.
```
      .03
4.7).141
     141
       0
```

1. .008
2. .09
3. .006
4. .007
5. .07
6. .0004
7. .056
8. .0006
9. .05

Page 45: Work for a Zero Remainder

A.
```
  3.4
5)17.0
  15
   20
   20
    0
```

2. .375

3.
```
  1.25
4)5.00
  4
  10
   8
   20
   20
    0
```

4. .875

5.
```
  1.75
4)7.00
  4
  30
  28
   20
   20
    0
```

6. .625
7. .25
8. .75
9. 1.25
10. 2.75
11. 3.25
12. 6.25

Page 46: Dividing by 10, 100, and 1,000

1. 63.53	**6.** 9.345	**11.** .1962
2. .51	**7.** 4.39	**12.** .841
3. .208	**8.** 1.674	**13.** .3054
4. 4.76	**9.** .41	**14.** .623
5. 42.5	**10.** .9281	**15.** .480

Page 47: Mastering the Skills

1. .04	**5.** .5	**9.** 150
2. 2.3	**6.** 1.01	**10.** .004
3. .2	**7.** 9	**11.** 60
4. .7	**8.** 80	**12.** .4

Page 48: Practice Your Skills

1. 4
2. 1230
3. .081
4. 34.7
5. 3.4
6. 14.7
7. 1.4605
8. 1050
9. .08
10. 3.0545

Page 49: Division Review

1. a) 24
   b) 3
   c) 8
2. .07
3. .006
4. 9.2
5. 4
6. 23.9
7. .26
8. 31
9. 56.8
10. 130
11. 12,000
12. .02
13. 1.25
14. 8.36
15. 4.26

Page 50: Use All Operations

1. 9.53
2. 73
3. 1,047
4. 148
5. 368
6. .18
7. 82
8. 64.952
9. 297.18
10. 45.37
11. 798.39
12. 558.44

Page 51: Finding Averages

A. 86.6°

1. $1.77
2. 21.63 miles per gallon
3. 85.1°
4. 37.19

Page 52: Does the Answer Make Sense?

1. $14.10 ÷ 6 = $2.35; 2.35
2. $28.70 ÷ 7 = $4.10; 4.10
3. $2.36 ÷ $.79 = 3; 3
4. $18.25 ÷ $3.65 = 5; 5
5. $87.50 ÷ 12.50 = 7; 7
6. $35.55 ÷ 4.5 = $7.90; 7.90

Page 53: Choose to Multiply or Divide

A. $3.24

B. $.54

1. divide: You need to find one part of a total.
2. multiply: You need to find the total cost of several rolls.
3. divide: You need to find one part of a total.
4. divide: You need to find one part of a total.
5. multiply: You need to find the total cost of several gallons.
6. multiply: You need to find the total cost of several pounds.

Page 54: Decide to Multiply or Divide

1. 52.3 × 3 = 156.9; 156.9
2. $23.87 ÷ 13.5 = $1.77; 1.77
3. 123.2 ÷ 2.2 = 56; 56
4. $.95 × 4 = $3.80; 3.80
5. $57.75 ÷ $8.25 = 7; 7
6. $42.38 × 3 = $12.14; 12.14
7. $88.16 × 3 = $264.48; 264.48
8. $9.95 ÷ $1.99 = 5; 5

Page 55: Think It Through

Answers should be similar to these.

1. a) How much will one belt cost?
   b) How much will 3 belts cost?
2. a) How many sacks of apples are there?
   b) What will 6 sacks of apples cost?
3. a) What will one gallon of paint cost?
   b) What will 3 gallons of paint cost?
4. a) What will 1.25 pounds of beef cost?
   b) What will 5 pounds of beef cost?
5. a) How many people are in each van?
   b) How much will 32 people pay altogether?
6. a) How many boys and girls bought tickets?
   b) How much did the 40 boys and girls pay for their tickets?

Page 56: Write a Question

Questions and answers will vary.

Page 57: Using Symbols

1. $29.46 < 30$
2. $32.04 > 32$
3. $83 > 82$
4. $821 = 821$
5. $574 > 486.2$
6. $365.04 < 367.48$
7. $161 = 161$
8. $30 < 30.99$

Page 58: Choose the Operation

1. multiplication
2. division
3. subtraction
4. addition
5. addition
6. multiplication
7. subtraction
8. addition
9. multiplication
10. division

Page 59: Apply Your Skills

1. $9.5 \times 5 = 47.5$; 47.5
2. $\$3.68 + \$4.65 = \$8.33$; 8.33
3. $\$14.97 \div 3 = \$4.99$; 4.99
4. $\$984.16 - \$544.46 = \$439.70$; 439.70
5. $\$479.55 \div 5 = \$95.51$; 95.51
6. $21.5 \times 7 = 150.5$; 150.5

Page 60: Mixed Practice

1. $\$91.16 + \$100.64 = \$191.80$; 191.80
2. $\$18.75 \div \$6.25 = 3$; 3
3. $\$47.18 + \$4.13 = \$51.31$; 51.31
4. $\$5.67 \div \$1.89 = 3$; 3
5. $\$16.75 + \$5.75 = \$22.50$; 22.50
6. $18.75 \times 9 = 168.75$; 168.75

Page 61: Review the Operations

1. C
2. A
3. B
4. D
5. \$43.47
6. \$43.47
7. $\$125.92 > \$82.45$
8. $\$82.45 < \$125.92$
9. $\$89.82 \div 6 = \$14.97$
10. $\$1.35 \times 15 = \$20.25$

Page 62: Two-Step Story Problems

1. a) $\$235.64 - \$50.00 = \$185.64$
   b) $\$185.64 \div 6 = \$30.94$
2. a) \$350.00
   b) \$324.25
3. a) \$32.15
   b) \$7.85
4. a) \$50.55
   b) \$9.45
5. a) \$2.50
   b) \$3.25

Page 63: More Two-Step Problems

1. $150.72
2. $560.60
3. $2.14
4. $23.84
5. 150
6. $12.00
7. a) $.49
   b) $3.43

Page 64: Multistep Word Problems

Answers should be similar to these.

1. Question 1: How much will the adult tickets cost? $27.75
   Question 2: How much will the children's tickets cost? $19.80
   Question 3: How much will the tickets cost altogether? $27.75 + $19.80 = $47.55
2. Question 1: How much did the two pairs of slacks cost? $65.00
   Question 2: How much did the two sport coats cost? $311.70
   Question 3: How much did he pay altogether? $65.00 + $311.70 + $32.96 = $409.66
3. Question 1: How much did she pay for the ham? $11.97
   Question 2: What was her total bill? $15.72
   Question 3: How much change did she get back from $20? $20.00 – $15.72 = $4.28
4. Question 1: How many hours a week does he work? 32
   Question 2: How much does he earn in one week? $296.00
   Question 3: How long will it take him to earn $1,480? $1,480.00 ÷ $296.00 = 5 weeks

Page 65: Division Problem-Solving Review

1. $4.02
2. 45.125
3. multiply $17.52
4. divide $12.30
5. a) How much are 18 gallons of gas?
   b) How much is 1 gallon of gas?
6. a) $.30
   b) $3.60
7. addition; $9.00
8. multiply; $50.00

Page 66: Weekly Paycheck Stub

1. $678.98
2. Federal $93.99
   FICA $45.39
   State $27.15
   Total $166.53
3. Medical $12.09
   Union Dues $2.55
   Others $0
   Total $14.64
4. $166.53 + $14.64 = $181.17
5. $678.98 – $181.17 = $497.81

Page 67: Gross Pay and Net Pay

1.

Name	Week Ending		Gross Pay	Net Pay	
Millie Sanchez	12/31/04		$156.00	$117.06	
Tax Deductions			Optional Deductions		
Federal	FICA	State	Medical	Union Dues	Others
$23.40	$9.30	$6.24			

2. $38.94
3. $117.06
4. a) $305.25
   b) $224.20
5. a) $352.00
   b) $266.75
6. $12.55

Page 68: Unit Pricing

1. $.22 per ounce
2. $.16 per foot
3. $.29 per ounce
4. $.21 per ounce
5. $1.12 per pound
6. $.08 per piece

Page 69: Find the Best Buy

1. Jumbo
2. Jumbo Size / Regular Size

```
   Jumbo Size        Regular Size
       $0.22              $0.18
   24)$5.28           18)$3.24
       4 8                1 8
         48               1 44
         48               1 44
          0                  0
```

3. $.22 jumbo $.18 regular
4. Regular
5. Answers will vary.

Page 70: Comparison Shopping

1. 60-foot special
2. 12-pack
3. Store A
4. 26 oz for $5.12

Page 71: Sales Tax

1. $.06
2. $.12
3. $.18
4. $.03
5. $.21
6. $.04
7. $.19
8. $.07
9. $.11
10. $.16
11. a) $3.34
    b) $.21
    c) $3.55
12. a) $3.29
    b) $.21
    c) $3.50
13. a) $2.70
    b) $.17
    c) $2.87

Page 72: Eating Out

1. $12.68
2. $.64
3. a) $4.50
   b) $3.92
   c) $4.30
   d) $12.72
   e) $13.29
   f) $6.71
4. $.36
5. Hamburger, milk shake, fries
6. a) $24.75
   b) $4.90
   c) $8.60
   d) $4.50
   e) $42.75
   f) $45.14
   g) $4.86

Page 73: Real-Life Problems

1. a) $2.97
   b) $11.88
2. a) $10.20
   b) $11.59
   c) $4.75
   d) $26.54
3. $9.10
4. a) $.39
   b) $.43
   c) $.41
   d) $.42
5. $86.18
6. $9.35

Page 74: Life-Skills Math Review

1. $.22
2. $5.11
3. a 16-ounce box of cereal for $3.35
4. $312.25
5. $1.12
6. $3.12
7. $27.60
8. $403.60

Page 75: Cumulative Review

1. a) .01292
   b) 3,865.8816
2. a) 745
   b) 633.5
3. a) 5.68
   b) 1.6
4. a) 2.3
   b) 6
5. $20.51

Page 76: Posttest

1. .45
2. .054
3. 33.75
4. $6.54
5. 130.9 miles
6. .004
7. 1.5275
8. 92
9. .00371
10. .83
11. .0052
12. 405
13. $9.35
14. $2.60
15. .066
16. 4 pounds
17. 62,500 pounds
18. 280
19. 54.8 mph
20. $32.00

Posttest Evaluation Chart

If a student misses one or more problems in a skill area, assign a review of the practice pages for that skill.

Skill Area	Posttest Problem #	Skill Section	Review Page
Multiplication	2, 3, 7, 9, 12, 18	7–22	23, 50
Division	1, 6, 8, 10, 11, 15	32–48	49, 50
Multiplication Problem Solving	5, 13, 17, 20	24–30	31
Division Problem Solving	4, 14, 16, 19	51–64	65
Life-Skills Math	All	66–73	74

FRACTIONS
The Meaning of Fractions

The format for teaching fractions is similar to that used for whole numbers. You should develop an understanding of fractions by proceeding from the concrete to the abstract. There are basically three different models that can be used to develop an understanding of fractions: 1) regions, 2) sets, and 3) number lines.

The more emphasis placed on the understanding of fractions, the less time that will be needed to develop an understanding of applying operations to fractions. Students are less likely to become frustrated when skills are repeated using different activities. This results in a longer attention span, thus increasing master of the skills.

Student Glossary

Acquainting students with definitions of key math terms and life-skills concepts will enhance their mastery of the materials. Below are words defined in the student text. A glossary is provided at the end of the student text and on page 243 of the *Teacher's Resource Guide and Answer Key.*

denominator	hour	proper fraction
divisible	improper fraction	remainder
equivalent fraction	mile	simplify (reduce)
factor	mixed number	sum
fraction	numerator	symbol
gallon	pound	whole number
greatest common factor	product	

STUDENT PAGE
7
Identifying Fractions

Understanding a Fraction

Divide an object into equal parts. Have students count the parts. Write the number of parts under the fraction bar and tell students that this is the **denominator** of a fraction. Then show students that the **numerator** tells how many equal parts are shaded or used out of the whole object.

EXAMPLE

Have students count equal parts together as you point at regions of a total figure.

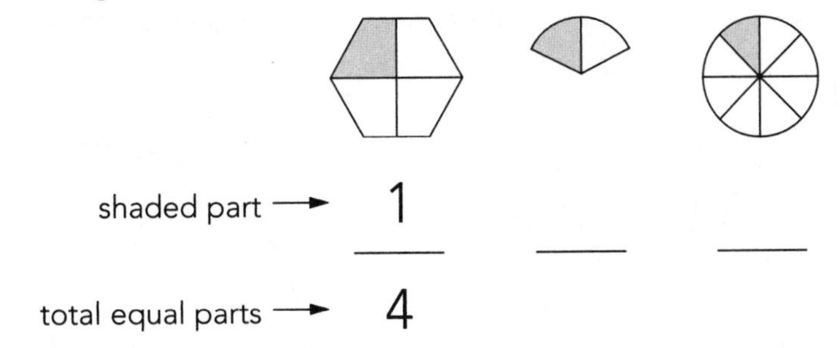

SAMPLE QUESTIONS

How many equal parts are there?

How many equal parts are shaded?

What does the top number (numerator) of the fraction tell us?

What does the bottom number (denominator) tell us?

EXAMPLE

Have students shade one of the equal parts for different pictorial models and name the fraction.

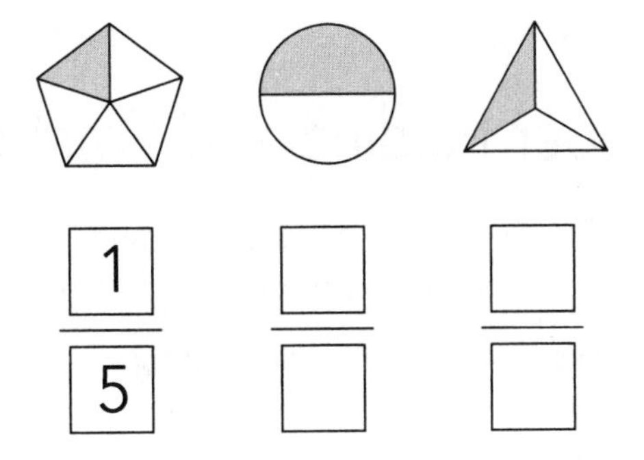

EXAMPLE

Have students name fractions with a numerator of 1 ($\frac{1}{4}$, $\frac{1}{9}$, etc.). Then continue with drawings of fractions with numerators larger than 1 using the same questioning process.

STUDENT PAGE

Identifying Fractions

Fraction Wheel

1 Cut out an enlarged copy of the two circles.

2 Cut along AB and CD.

3 Fit the wheels together by inserting CD into AB.

4 Call out fractions and have students show each fraction by rotating the shaded circle.

EXAMPLE Students insert the two wheels with the dark wheel on top. Have students rotate the wheels so that CD lines up with $\frac{1}{4}$. They will see $\frac{1}{4}$ of the dark wheel.

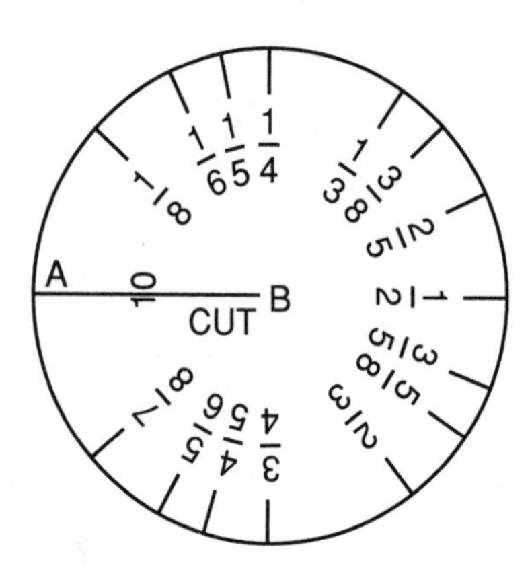

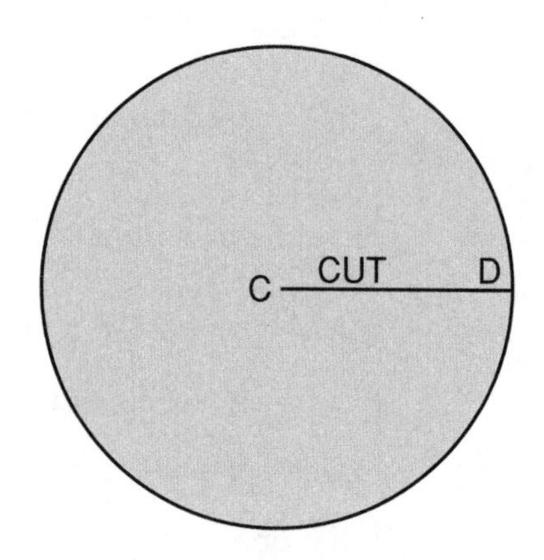

Alternate Activity

Have students compare fractions by using the wheel to determine which is greater, $\frac{3}{4}$ or $\frac{3}{8}$.

STUDENT PAGE

Identifying Fractions

Paper Folding and Fractions

1 Fold a piece of paper in half and cut with scissors. Demonstrate that two halves equal one whole or one of something.

2 Have students fold and cut (or sketch and shade) fractions using different figures. Start with rectangles. Continually check students' understanding to ensure that they are dividing the total of a figure into equal parts.

3 Compare and discuss drawings of $\frac{1}{2}$, $\frac{1}{4}$, $\frac{1}{3}$, $\frac{3}{4}$, $\frac{2}{3}$, etc.

SAMPLE QUESTIONS

Is $\frac{1}{4}$ or $\frac{1}{3}$ larger?

Which is smaller: $\frac{3}{4}$ or $\frac{2}{3}$?

STUDENT PAGE
10
Identifying Fractions

Fractions Outside the Classroom

Have students find and clip fractions as they appear in a merchandise flier or a single day's newspaper. They can paste together a collage displaying the fractions.

Alternate Activity

Have the class create a number line—moving from smallest to largest—using the fractions students clipped from the newspaper.

STUDENT PAGE
10
Identifying Fractions

Equal Parts

Give students a drawing of an equilateral triangle and have them divide it into 4 equal parts. Then tell them to shade three-fourths.

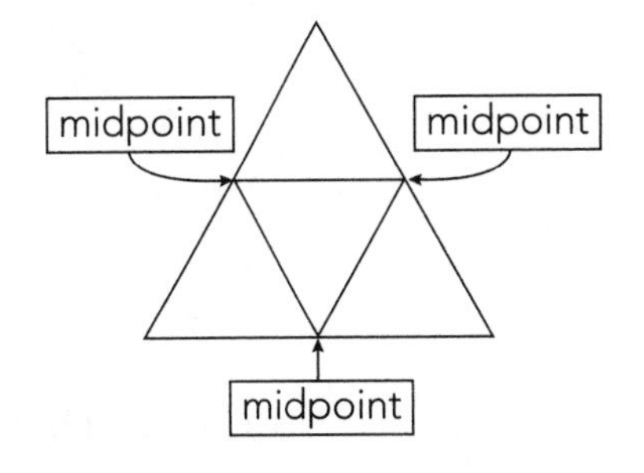

This activity will challenge many students. Some will divide the triangle into shapes that are not equal parts. This is a good chance to emphasize that the parts of the total must be divided into equal, or **congruent**, parts. You may have to provide a hint by introducing the concept of **midpoint**. Once students find the midpoint of each side of the triangle, the rest should be easy.

STUDENT PAGE
15
Identifying Fractions

Identify the Fractions and Whole Numbers

Write a scrambled list of **whole numbers, proper fractions, improper fractions,** and **mixed numbers** on the board. Students can draw and fill in a chart identifying each of the whole numbers and types of fractions.

Whole Numbers	Proper Fractions	Improper Fractions	Mixed Numbers
9	$\frac{4}{5}$	$\frac{7}{4}$	$2\frac{1}{2}$
___	___	___	___
___	___	___	___
___	___	___	___

STUDENT PAGE
17
Identifying Fractions

Using Sets

Students should develop an understanding of fractions by working with sets. Have students work with concrete manipulatives, such as pencils or clothespins.

$\frac{3}{6}$ { ⓧ ⓧ ⓧ / X X X }

Consider 3 separate objects out of a total of 6. We are now showing 3 out of 6 or $\frac{1}{2}$.

$\frac{6}{8}$ { ⓧ ⓧ ⓧ X / ⓧ ⓧ ⓧ X }

Consider 6 separate objects out of a total of 8. We are now showing 6 out of 8 or $\frac{6}{8}$.

Many other fractions can be demonstrated in the same way.

STUDENT PAGE
17
Identifying Fractions

Using Number Lines

Students should develop an understanding of fractions using number lines. Have students count the marks including the number 1. Then ask them to name the fraction for each line segment.

EXAMPLE

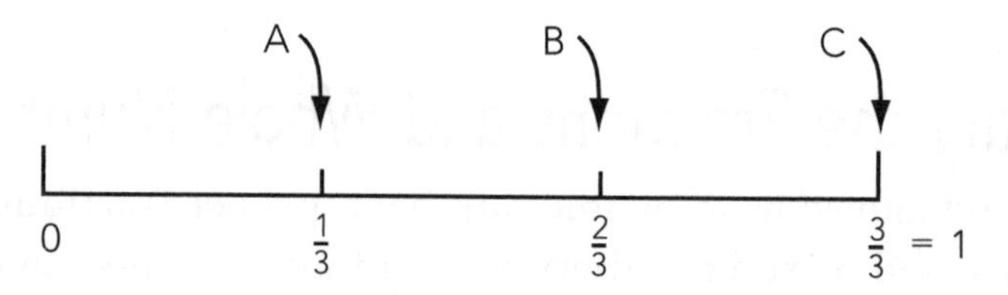

STUDENT PAGE
27
Equivalent Fractions

Folding Paper to See Equivalent Fractions

Fold paper into halves, fourths, eighths, and sixteenths and label the resulting parts. Demonstrate that $\frac{2}{2} = \frac{4}{4} = \frac{8}{8} = \frac{16}{16}$.

Compare and discuss equivalent fractions. $\frac{1}{2} = \frac{2}{4} = \frac{4}{8} = \frac{8}{16}$

$\frac{1}{2}$	$\frac{1}{2}$

$\frac{1}{4}$	$\frac{1}{4}$
$\frac{1}{4}$	$\frac{1}{4}$

$\frac{1}{8}$	$\frac{1}{8}$	$\frac{1}{8}$	$\frac{1}{8}$
$\frac{1}{8}$	$\frac{1}{8}$	$\frac{1}{8}$	$\frac{1}{8}$

$\frac{1}{16}$	$\frac{1}{16}$	$\frac{1}{16}$	$\frac{1}{16}$	$\frac{1}{16}$	$\frac{1}{16}$	$\frac{1}{16}$	$\frac{1}{16}$
$\frac{1}{16}$	$\frac{1}{16}$	$\frac{1}{16}$	$\frac{1}{16}$	$\frac{1}{16}$	$\frac{1}{16}$	$\frac{1}{16}$	$\frac{1}{16}$

STUDENT PAGE
27
Equivalent Fractions

Working with Sets to Show Equivalent Fractions

Demonstrate that $\frac{1}{2} = \frac{3}{6}$ and $\frac{3}{4} = \frac{9}{12}$ using sets.

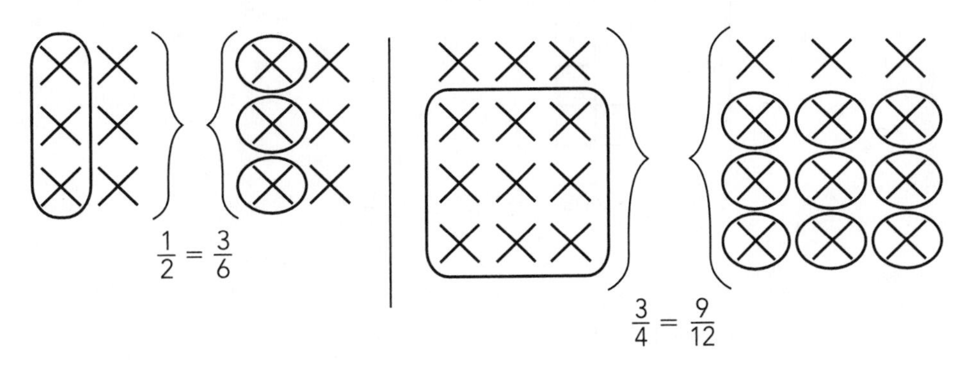

Students should observe and work with many other representations to help them identify these two important concepts:

1. A fraction can represent part of a whole.
2. A fraction can represent part of a set.

STUDENT PAGE
29
Equivalent Fractions

Shading Equivalent Fractions

By coloring and comparing equivalent regions, students can see that fractions can be renamed in different ways.

Have students shade each row of fractions with a different colored pencil.

EXAMPLES

A

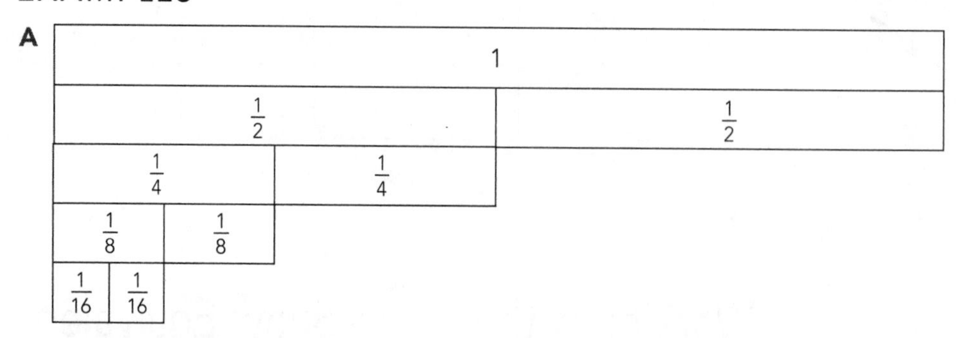

B

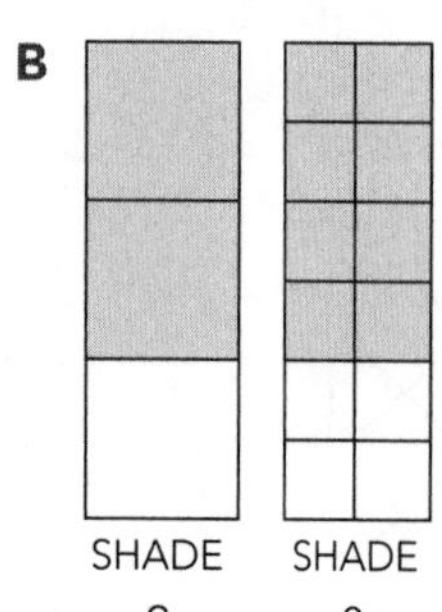

SHADE SHADE

$\frac{2}{3} = \frac{8}{12}$

C

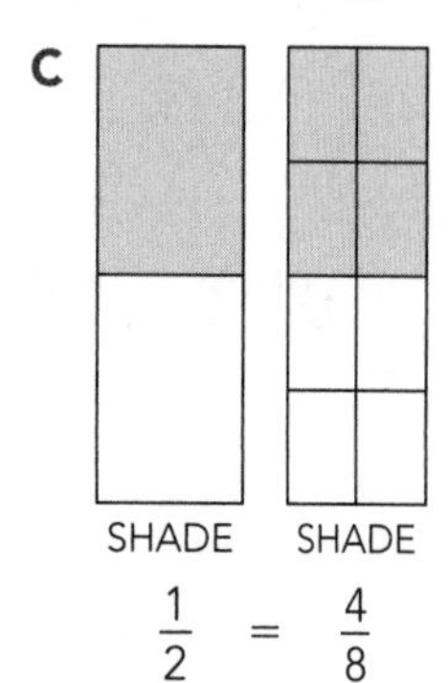

SHADE SHADE

$\frac{1}{2} = \frac{4}{8}$

STUDENT PAGE
36
Comparisons

Larger and Smaller Fractions

The number line is an excellent device for developing an understanding of the relationships between fractions. Students should learn that as they move to the right on the number line, whole numbers and fractions become larger.

EXAMPLES

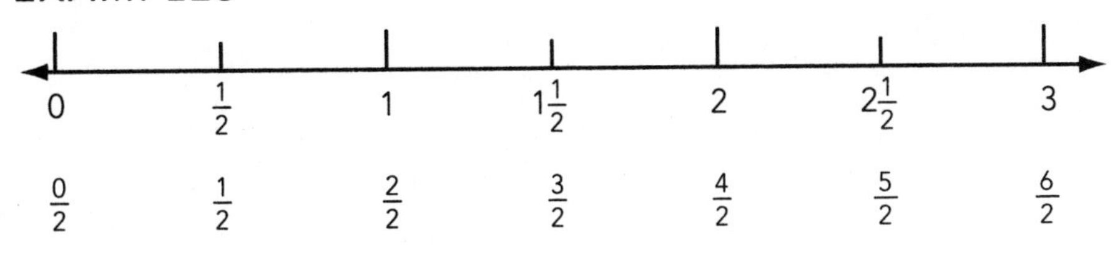

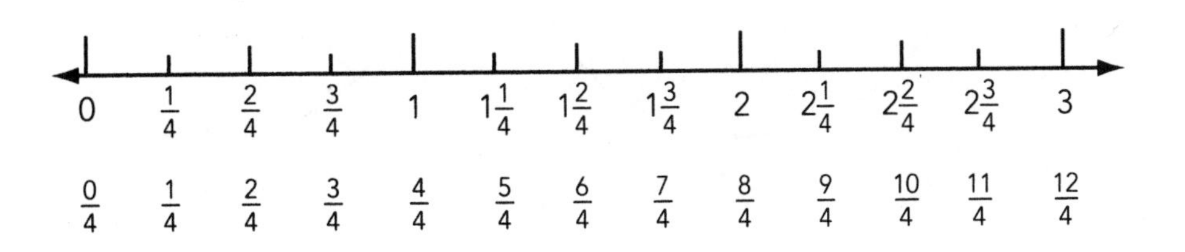

SAMPLE QUESTIONS

Which is larger: $\frac{5}{2}$ or 2?

Which of these is equal to $1\frac{3}{4}$: $\frac{9}{4}$ or $\frac{7}{4}$?

STUDENT PAGE
36
Comparisons

Estimating Fractions

Use two paper plates with contrasting colors. A cut should be made on both plates along the lines indicated. Interlock the two circles and rotate for different-size fractions.

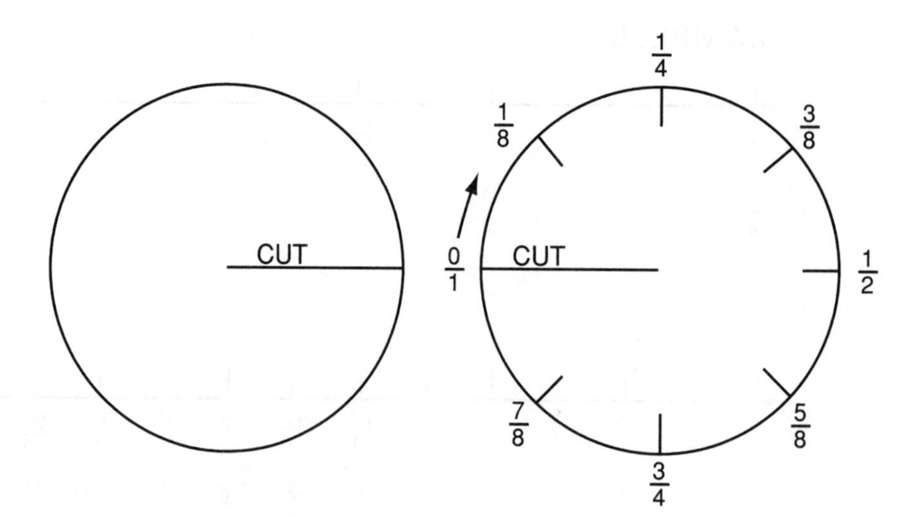

SAMPLE ACTIVITIES

If the fraction shown is less than $\frac{1}{2}$, show thumbs down.

If the fraction shown is greater than $\frac{1}{2}$, show thumbs up.

If the fraction shown is greater than $\frac{3}{4}$, show thumbs up.

If the fraction shown is less than $\frac{3}{4}$, show thumbs down.

Alternate Activity

Using the circles made from paper plates, let your students complete expressions using greater than (>) and less than (<) symbols.

$\frac{1}{4}$ (<) $\frac{3}{8}$ $\frac{3}{4}$ ◯ $\frac{7}{8}$

$\frac{5}{8}$ ◯ $\frac{1}{2}$ $\frac{1}{4}$ ◯ $\frac{1}{8}$

STUDENT PAGE
45
Divisibility and Common Fractions

Sieve of Eratosthenes

Another way of reinforcing divisibility rules is by working with numbers. The following arrangement of number is called the Sieve of Eratosthenes because it was developed by the Greek geographer Eratosthenes and it looks like a sieve.

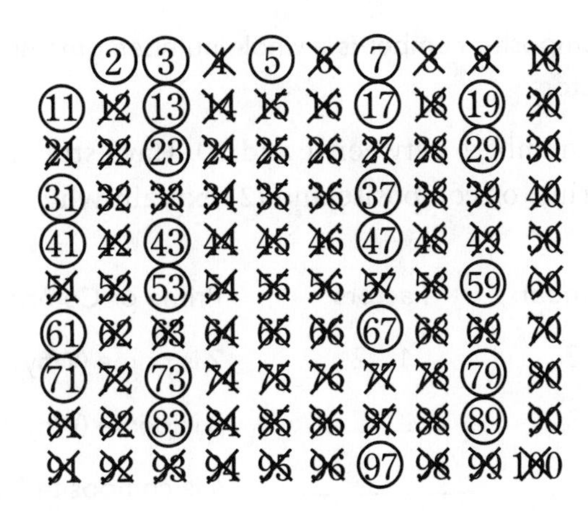

The sieve can be used to point out simple rules for divisibility for the numbers 2, 3, 5, and 7.

Follow the pattern:

1. Circle number 2 and cross out all numbers divisible by 2.
2. Circle number 3 and cross out all numbers divisible by 3 that have not already been crossed out.
3. Circle number 5 and cross out all numbers divisible by 5 that have not already been crossed out.
4. Circle number 7 and cross out all numbers divisible by 7 that have not already been crossed out.
5. Have students circle all remaining numbers that have not been crossed out.
6. Have students make a list of all numbers that are not crossed out. (Answer: 2, 3, 5, 7, 11, 13, 17, 19, 23, 29, 31, 37, 41, 43, 47, 53, 59, 61, 67, 71, 73, 79, 83, 89, 97)

Teach students that these whole numbers are greater than 1 and have two and only two whole-number factors: 1 and the number itself. These numbers are called **prime numbers.**

The whole numbers greater than 1 that have more than two whole-number factors are called **composite numbers.**

Ask the students what they notice about all prime numbers except 2? (Answer: They are odd numbers.)

STUDENT PAGE
47
Divisibility and Common Fractions

Prime and Composite Numbers

Every whole number greater than 1 can be named by at least two factors: 1 and the number itself.

A **prime number** is a number greater than 1 that has only two factors: 1 and the number itself.

A **composite number** is a whole number, greater than 1, with more than 2 factors.

List numbers between 1 and 20. Have students (1) tell whether the number is prime or composite and (2) explain why.

Number	Factors	Prime or Composite
2	1 2	2 is prime (Why?)
3	___ ___	3 is prime (Why?)
4	___ ___ ___	4 is composite (Why?)
5	___ ___	5 is ___________ (Why?)

STUDENT PAGE
47
Divisibility and Common Fractions

Factor Trees

Every composite number can be expressed as the product of prime factors. Have students complete the **factor trees** for different composite numbers.

Students should notice that every number in the last row is a prime number. Remember that 1 is neither prime nor composite.

First, start some factor trees so the student can fill in the answers without a lot of questions.

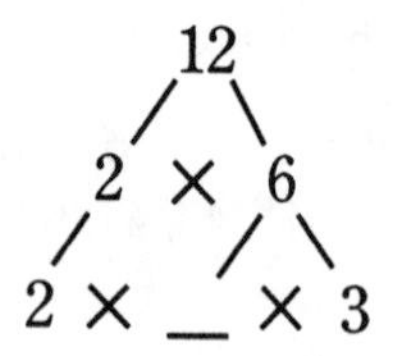

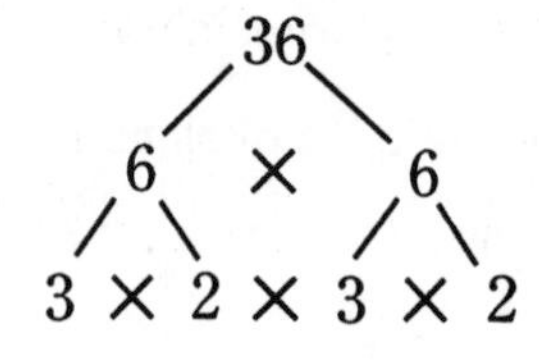

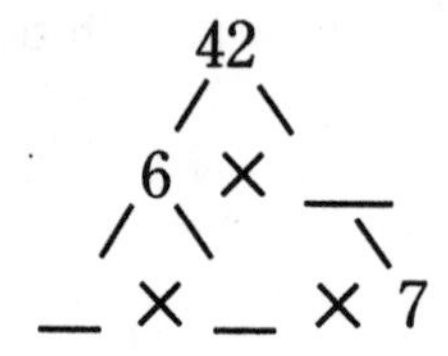

STUDENT PAGE
54
Divisibility and Common Fractions

Mentally Find Equivalent Fractions

Scramble a series of fractions in a square and have your students list as many pairs of equivalent fractions as they can find in 1 minute.

EXAMPLE

$\frac{1}{2}$	$\frac{2}{3}$	$\frac{6}{9}$	
	$\frac{2}{20}$	$\frac{1}{10}$	$\frac{1}{5}$
$\frac{3}{6}$	$\frac{5}{25}$	$\frac{5}{10}$	

STUDENT PAGE
67
Life Skills Math

Using Rulers

Ask questions about a ruler marked off in $\frac{1}{16}$-inch divisions. Let students use a ruler to answer questions until they become familiar with the divisions.

SAMPLE QUESTIONS

How many sixteenths of an inch are there in 2 inches? (Answer: 32)

How many quarters of an inch are there in $1\frac{1}{2}$ inches? (Answer: 6)

How many eighths of an inch are there in $2\frac{3}{4}$ inches? (Answer: 22)

STUDENT PAGE
67
Life Skills Math

Measurement Fractions

Using a ruler, have students label the drawings to show fractions of a foot or yard.

3 feet = 1 yard

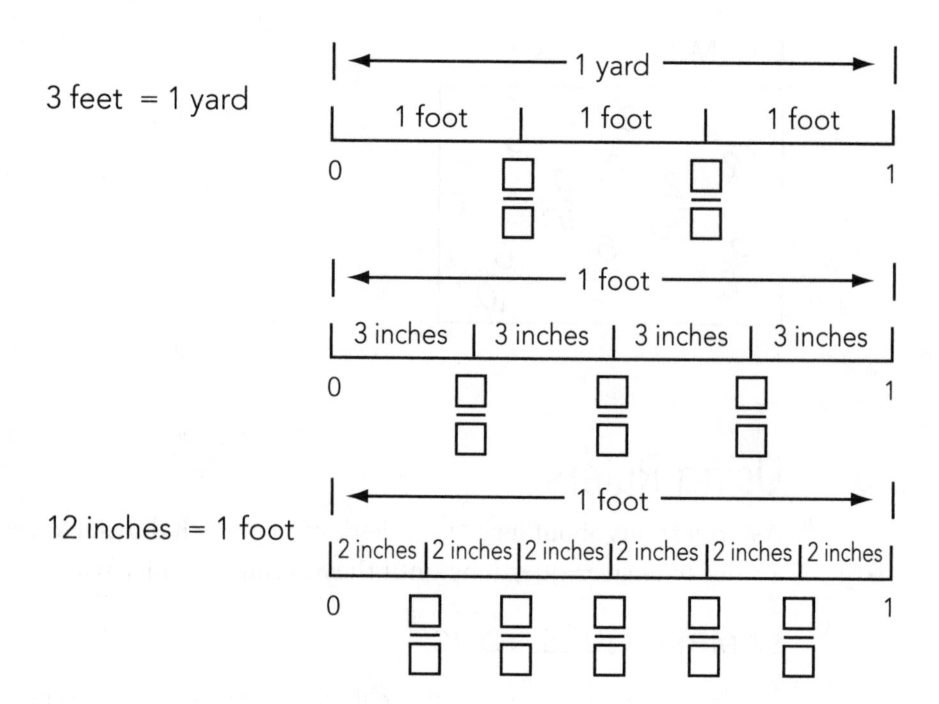

12 inches = 1 foot

STUDENT PAGE
68
Life Skills Math

Real-Life Fractions

Money and clocks are excellent concrete models of fraction relationships.

Have students think about $\frac{1}{2}$ of a dollar and explain how much money it represents. Do the same with $\frac{1}{100}$, $\frac{1}{20}$, $\frac{1}{10}$, $\frac{1}{4}$, $\frac{1}{5}$, $\frac{2}{5}$, $\frac{3}{4}$, etc., of a dollar.

Alternate Activity

Have students think about $\frac{1}{2}$ of an hour and explain how many minutes it represents. Do the same with $\frac{1}{4}$, $\frac{1}{5}$, $\frac{1}{6}$, $\frac{1}{10}$, $\frac{1}{20}$, $\frac{3}{4}$, $\frac{2}{5}$, etc., of an hour.

Page 4: Pretest

1. $\frac{2}{4}$ and $\frac{1}{2}$
2. $\frac{1}{6}$
3. $\frac{1}{7}$
4. $\frac{5}{6}$
5. $\frac{6}{12}$
6. $\frac{10}{15}$
7. $\frac{16}{18}$
8. $\frac{21}{28}$
9. $\frac{3}{10}$
10. $\frac{2}{3}$
11. $\frac{3}{4}$
12. $\frac{4}{9}$
13. $\frac{1}{4}$
14. $\frac{1}{3}$
15. $\frac{3}{5}$
16. $\frac{7}{10}$
17. .375
18. $.16\frac{2}{3}$
19. 1.75
20. 1.1

Pretest Evaluation Chart

If a student misses any problems in a skill area, assign the practice pages for that skill. However, students may need to complete all practice pages to reinforce areas of weakness.

Skill Area	Pretest Problem #	Skill Section	Review Page
Identifying Fractions	1, 2	7–19	20
Comparing Fractions	3, 4	21–25	26
Equivalent Fractions	5, 6, 7, 8	27–33	34
Comparisons	9, 10, 11, 12	35–39	40
Simplifying Fractions	13, 14, 15, 16	41–57	58
Fractions and Decimals	17, 18, 19, 20	59–62	63
Life-Skills Math	All	64–73	72, 74

Page 7: Denominators

1. $\frac{1}{4}$
2. $\frac{1}{2}$
3. $\frac{1}{8}$
4. $\frac{1}{3}$
5. $\frac{1}{6}$
6. $\frac{1}{5}$
7. $\frac{1}{8}$
8. $\frac{1}{2}$
9. $\frac{1}{8}$
10. $\frac{1}{3}$
11. $\frac{1}{4}$
12. $\frac{1}{4}$
13. $\frac{1}{5}$
14. $\frac{1}{2}$
15. $\frac{1}{3}$
16. $\frac{1}{2}$
17. $\frac{1}{4}$
18. $\frac{1}{4}$
19. $\frac{1}{10}$
20. $\frac{1}{6}$
21. $\frac{1}{4}$
22. $\frac{1}{3}$

Page 8: Numerators

1. $\frac{3}{4}$
2. $\frac{1}{2}$
3. $\frac{2}{5}$
4. $\frac{1}{4}$
5. $\frac{2}{4}$ or $\frac{1}{2}$
6. $\frac{2}{3}$
7. $\frac{3}{4}$
8. $\frac{4}{6}$ or $\frac{2}{3}$
9. $\frac{4}{8}$ or $\frac{1}{2}$
10. $\frac{1}{3}$
11. $\frac{3}{5}$
12. $\frac{3}{6}$ or $\frac{1}{2}$
13. $\frac{3}{4}$
14. $\frac{4}{5}$
15. $\frac{3}{4}$

Page 9: Fractions with Numerators of 1

1. $\frac{1}{5}$ = D **4.** $\frac{1}{7}$ = A **7.** $\frac{1}{4}$ = B

2. $\frac{1}{8}$ = G **5.** $\frac{1}{10}$ = I **8.** $\frac{1}{6}$ = E

3. $\frac{1}{9}$ = H **6.** $\frac{1}{3}$ = F **9.** $\frac{1}{2}$ = C

10. $\frac{1}{5}$ **11.** $\frac{1}{6}$ **12.** $\frac{1}{3}$ **13.** $\frac{1}{7}$

14. $\frac{1}{2}$

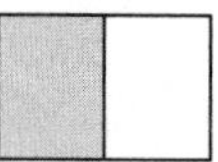

15. $\frac{1}{8}$

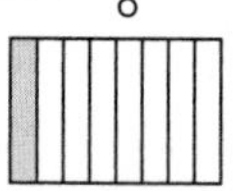

16. $\frac{1}{4}$

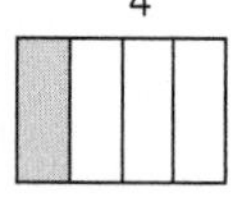

17. $\frac{1}{9}$

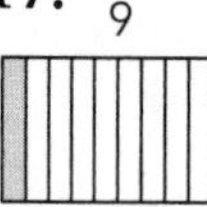

Page 10: Match the Fractions

1. $\frac{2}{3}$ = A **6.** $\frac{2}{4}$ = B

2. $\frac{3}{6}$ = E **7.** $\frac{5}{8}$ = G

3. $\frac{1}{4}$ = F **8.** $\frac{3}{4}$ = J

4. $\frac{5}{6}$ = I **9.** $\frac{1}{3}$ = D

5. $\frac{3}{5}$ = C **10.** $\frac{1}{2}$ = H

11. $\frac{2}{5}$

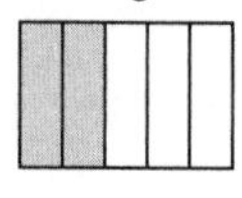

12. $\frac{3}{4}$

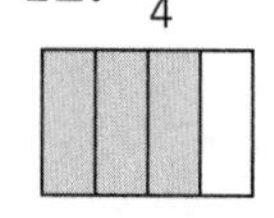

13. $\frac{1}{6}$

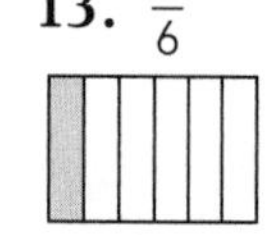

14. $\frac{3}{5}$

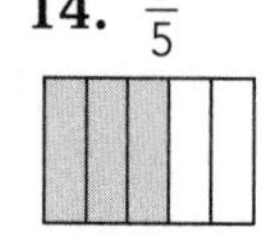

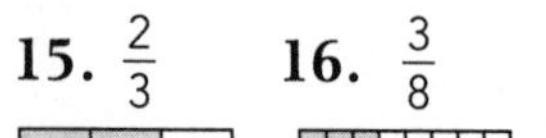

15. $\frac{2}{3}$

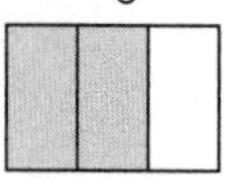

16. $\frac{3}{8}$

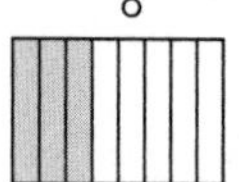

17. $\frac{1}{2}$

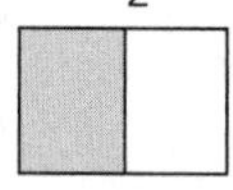

18. $\frac{3}{7}$

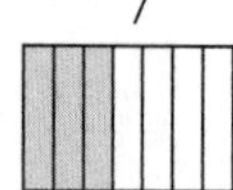

Page 11: Fractions Larger than 1

1. $\frac{7}{4}$ **3.** $\frac{5}{2}$ **5.** $\frac{5}{4}$

2. $\frac{6}{5}$ **4.** $\frac{4}{2}$ **6.** $\frac{7}{3}$

7. $\frac{3}{2}$

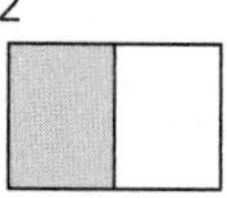

9. $\frac{6}{3}$

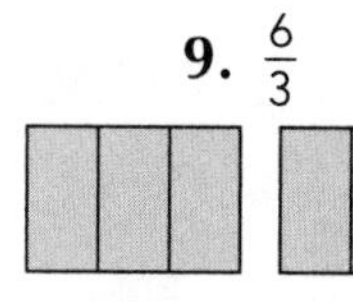

8. $\frac{5}{4}$

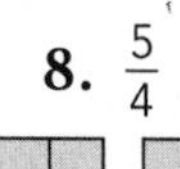

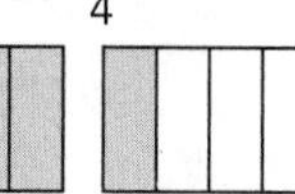

10. $\frac{7}{5}$

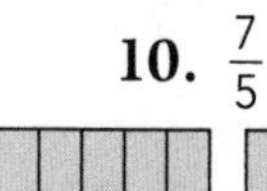

Page 12: Mixed Numbers

1. $2\frac{1}{2}$ **5.** $2\frac{5}{7}$ **9.** $3\frac{2}{3}$

2. $1\frac{2}{3}$ **6.** $5\frac{2}{9}$ **10.** $6\frac{3}{5}$

3. $1\frac{7}{10}$ **7.** $1\frac{7}{8}$ **11.** $2\frac{1}{6}$

4. $2\frac{3}{4}$ **8.** $7\frac{1}{2}$ **12.** $1\frac{1}{4}$

Page 13: Match the Mixed Numbers

1. $2\frac{2}{3}$ = C **5.** $1\frac{2}{3}$ = D

2. $1\frac{3}{4}$ = G **6.** $2\frac{1}{4}$ = E

3. $1\frac{1}{4}$ = A **7.** $2\frac{1}{2}$ = B

4. $1\frac{2}{6}$ = H **8.** $1\frac{1}{2}$ = F

9. $1\frac{1}{4}$ One and one fourth

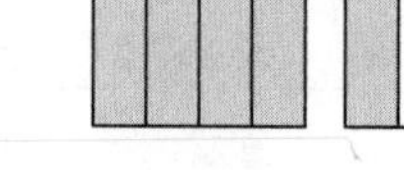

10. $2\frac{2}{5}$ Two and two fifths

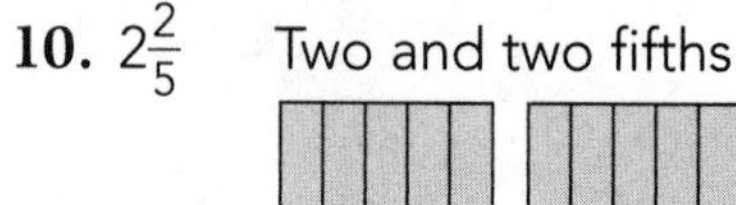

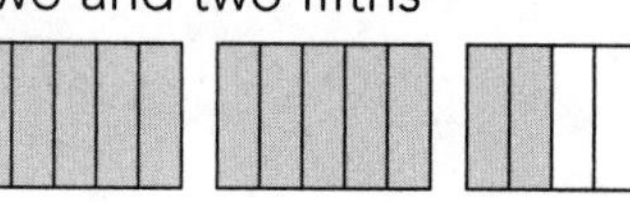

11. $2\frac{1}{3}$ Two and one third

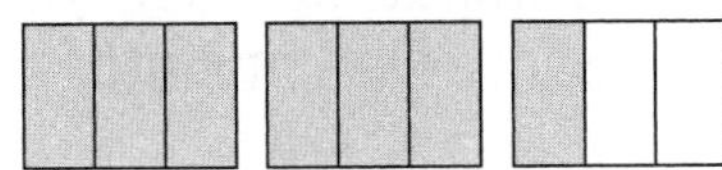

12. $2\frac{1}{5}$ Two and one fifth

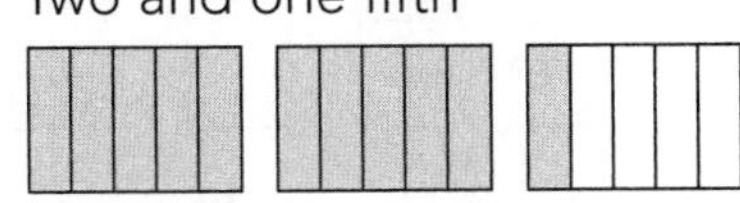

Page 14: Writing Fractions and Mixed Numbers

	Improper Fraction		Mixed Number
1.	$\frac{3}{2}$	=	$1\frac{1}{2}$
2.	$\frac{5}{3}$	=	$1\frac{2}{3}$
3.	$\frac{13}{5}$	=	$2\frac{3}{5}$
4.	$\frac{7}{2}$	=	$3\frac{1}{2}$
5.	$\frac{11}{4}$	=	$2\frac{3}{4}$
6.	$\frac{13}{6}$	=	$2\frac{1}{6}$

Page 15: Shade the Mixed Numbers

1. $1\frac{1}{2}$

2. $1\frac{3}{4}$

3. $1\frac{3}{5}$

4. $2\frac{3}{8}$

5. $2\frac{5}{6}$

Page 16: Write the Numbers

1. $1\frac{2}{6}$

2. $1\frac{4}{6}$

3. $\frac{2}{6}$

4. $1\frac{3}{6}$

5. $1\frac{1}{4}$

6. 2

7. $\frac{3}{4}$

8. $2\frac{2}{4}$

Page 17: Sets

1. H

2. A

3. B

4. D

5. E

6. F

7. C

8. G

9. $\frac{1}{2}$ of 4

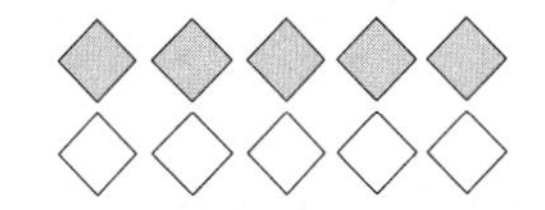

10. $\frac{1}{3}$ of 6

11. $\frac{1}{2}$ of 10

12. $\frac{1}{4}$ of 8

Page 18: More Sets

1. E

2. B

3. C

4. A

5. D

6. $\frac{3}{4}$ of 4

7. $\frac{2}{3}$ of 6

8. $\frac{4}{5}$ of 10

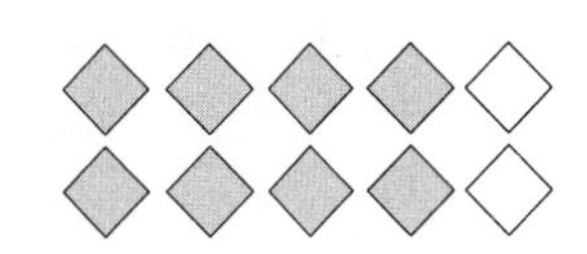

9. $\frac{3}{4}$ of 12

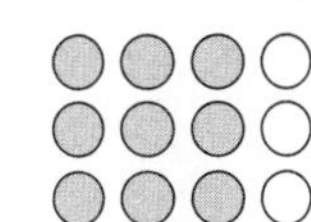

Page 19: Looking at Sets

Answers should be similar to these.

1. 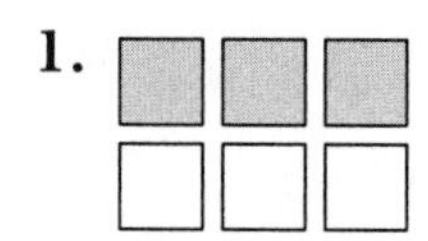

6. 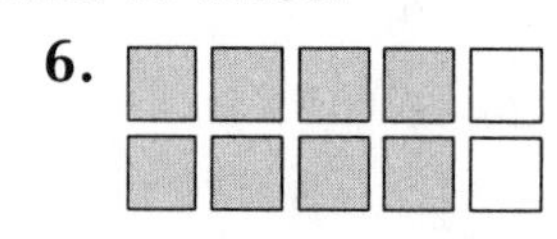

2. 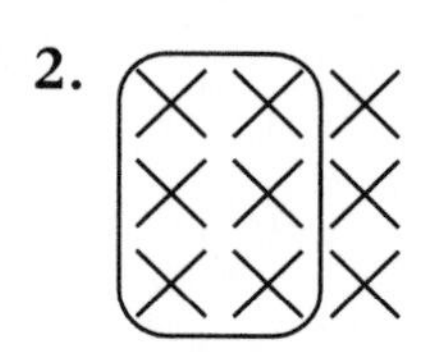

7. 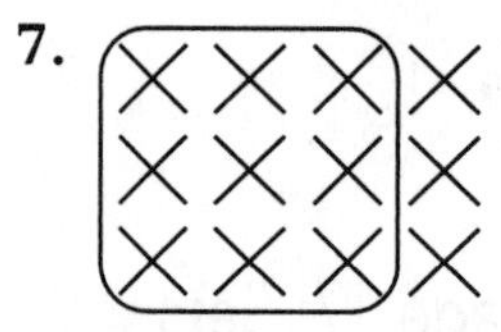

3. 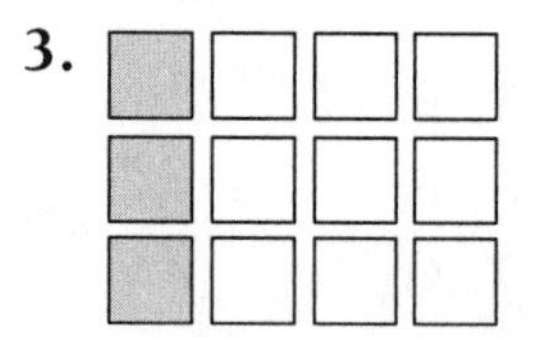

8. 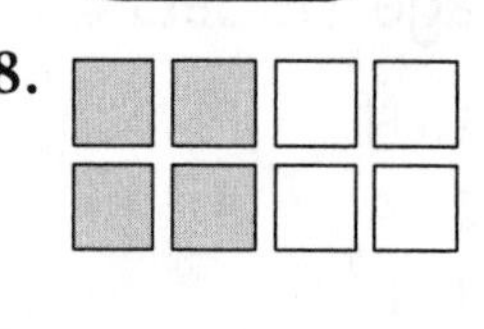

4. 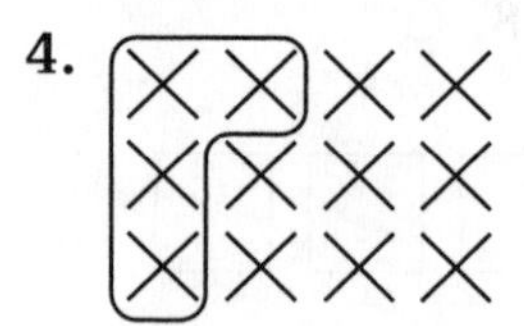

9. 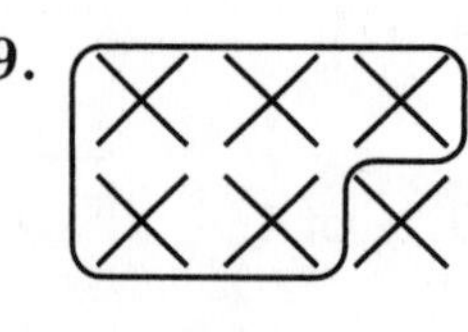

5. 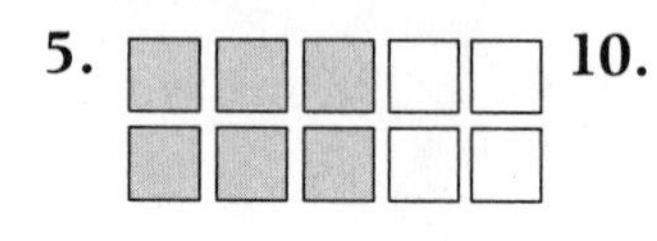

10. 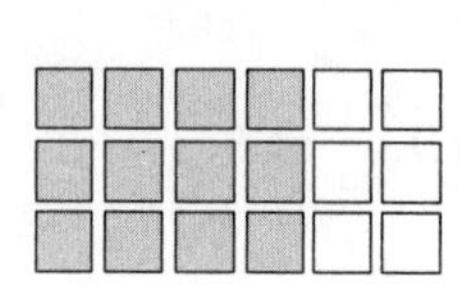

Page 20: Identifying Fractions Review

1. 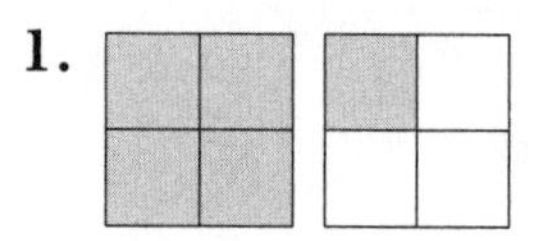

2. $\frac{2}{5}$

3. $2\frac{1}{3}$

4. $\frac{3}{8}$

5. 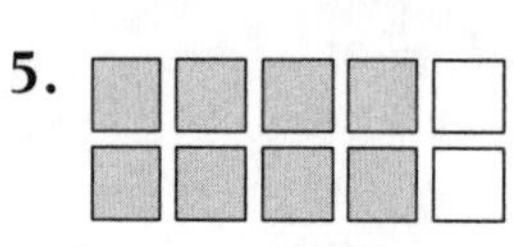

6. 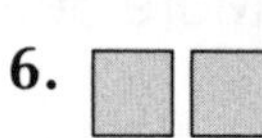

7. 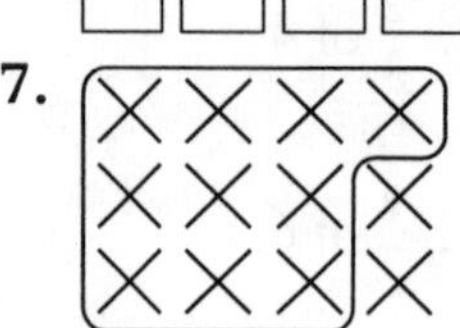

8. 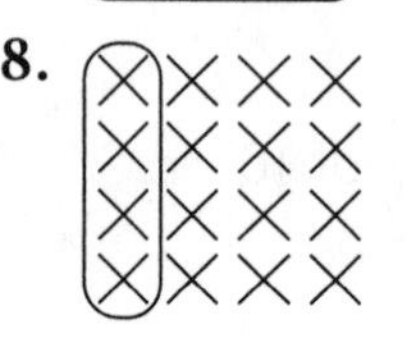

Page 21: Comparing Fractions with the Same Denominator

1. $\frac{2}{6}$ **2.** $\frac{5}{6}$ **3.** $\frac{1}{6}$ **4.** $\frac{3}{6}$ **5.** $\frac{6}{6}$ **6.** $\frac{4}{6}$

7. $\frac{1}{6}, \frac{2}{6}, \frac{3}{6}, \frac{4}{6}, \frac{5}{6}, \frac{6}{6}$

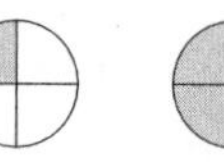
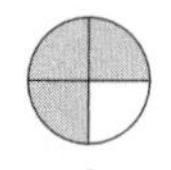

8. $\frac{1}{4}$ **9.** $\frac{3}{4}$ **10.** $\frac{2}{4}$ **11.** $\frac{0}{4}$ **12.** $\frac{4}{4}$

13. $\frac{0}{4}, \frac{1}{4}, \frac{2}{4}, \frac{3}{4}, \frac{4}{4}$

14. larger

15. smaller

Page 22: Comparing Fractions with the Same Numerator

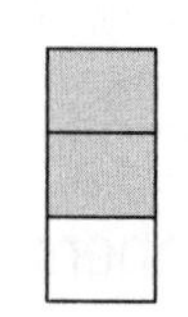
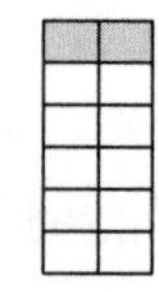
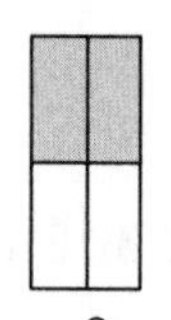
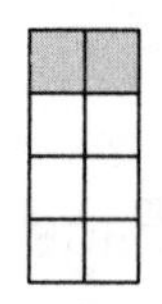

1. $\frac{2}{3}$ **2.** $\frac{2}{12}$ **3.** $\frac{2}{4}$ **4.** $\frac{2}{8}$

5. $\frac{2}{12}, \frac{2}{8}, \frac{2}{4}, \frac{2}{3}$

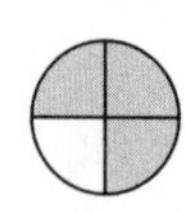

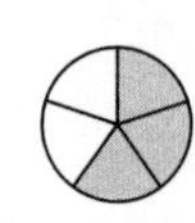
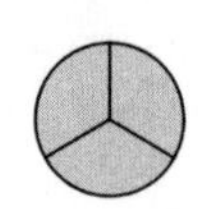

6. $\frac{3}{4}$ **7.** $\frac{3}{8}$ **8.** $\frac{3}{5}$ **9.** $\frac{3}{3}$

10. $\frac{3}{8}, \frac{3}{5}, \frac{3}{4}, \frac{3}{3}$

11. smaller

12. larger

Page 23: Shade and Compare

1. $\frac{2}{5}$
2. $\frac{1}{2}$
3. $\frac{1}{2}$
4. $\frac{2}{5}$
5. $\frac{5}{6}$
6. $\frac{1}{2}$
7. $\frac{5}{6}$
8. $\frac{1}{2}$
9. $\frac{4}{5}$
10. $\frac{1}{2}$
11. $\frac{4}{5}$
12. $\frac{1}{2}$
13. $\frac{1}{3}$
14. $\frac{1}{2}$
15. $\frac{1}{2}$
16. $\frac{1}{3}$
17. $\frac{5}{6}$
18. $\frac{4}{6}$
19. $\frac{5}{6}$
20. $\frac{4}{6}$
21. $\frac{3}{4}$
22. $\frac{6}{8}$
23. They are the same size.
24. $\frac{5}{3}$
25. $\frac{3}{2}$
26. $\frac{5}{3}$
27. $\frac{3}{2}$
28. $\frac{2}{3}$
29. $\frac{5}{6}$
30. $\frac{5}{6}$
31. $\frac{2}{3}$

Page 24: Order the Fractions

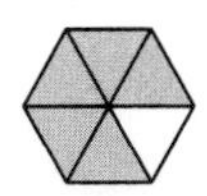

1. $\frac{5}{6}$

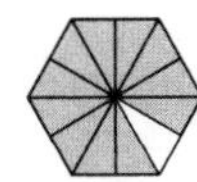

2. $\frac{11}{12}$

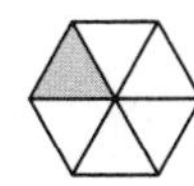

3. $\frac{1}{6}$

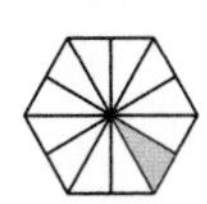

4. $\frac{1}{12}$

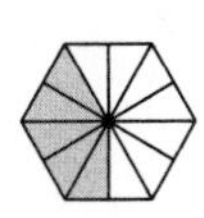

5. $\frac{5}{12}$

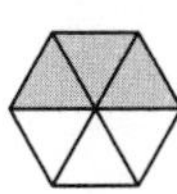

6. $\frac{3}{6}$

7. $\frac{1}{12}, \frac{1}{6}, \frac{5}{12}, \frac{3}{6}, \frac{5}{6}, \frac{11}{12}$

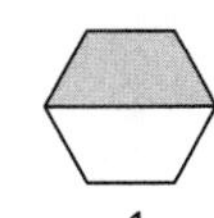

8. $\frac{1}{2}$

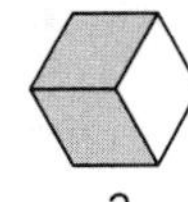

9. $\frac{2}{3}$

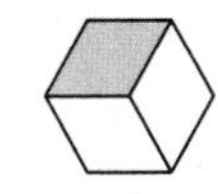

10. $\frac{1}{3}$

11. $\frac{1}{3}, \frac{1}{2}, \frac{2}{3}$

Page 25: Practice Helps

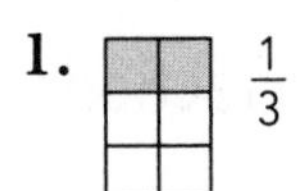

1. $\frac{1}{3}$
2. $\frac{1}{2}$
3. $\frac{5}{6}$
4. $\frac{1}{3}$ $\frac{1}{2}$ $\frac{5}{6}$

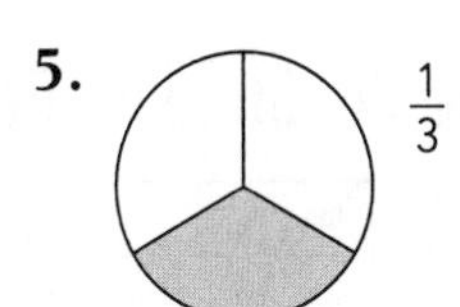

5. $\frac{1}{3}$

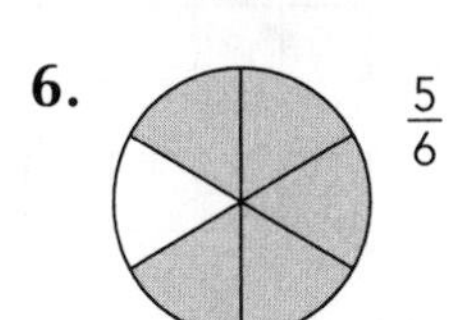

6. $\frac{5}{6}$

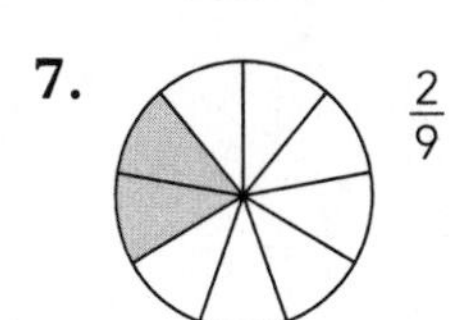

7. $\frac{2}{9}$
8. $\frac{2}{9}$ $\frac{1}{3}$ $\frac{5}{6}$

Page 26: Comparing Fractions Review

1. $\frac{4}{7}$
2. $\frac{5}{7}$
3. $\frac{5}{7}$
4. $\frac{4}{7}$
5. $\frac{1}{4}$
6. $\frac{3}{8}$
7. $\frac{3}{5}$
8. $\frac{3}{3}$
9. $\frac{1}{4}$ $\frac{3}{8}$ $\frac{3}{5}$ $\frac{3}{3}$
10. $\frac{4}{3}$
11. $\frac{3}{2}$
12. $\frac{3}{2}$
13. $\frac{4}{3}$

Page 27: What Are Equivalent Fractions?

1. yes
2. $\frac{1}{2}$
3. $\frac{2}{4}$
4. yes
5. yes
6. yes
7. $\frac{1}{2} = \frac{4}{8}$
8. $\frac{1}{4} = \frac{3}{12}$
9. $\frac{3}{4} = \frac{6}{8}$
10. $\frac{2}{3} = \frac{6}{9}$

Page 28: Shading Equivalent Fractions

1. $\frac{1}{2}$
2. $\frac{2}{4}$
3. $\frac{4}{8}$
4. yes
5. yes
6. $\frac{1}{2} = \frac{2}{4} = \frac{4}{8}$

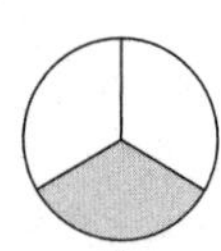

7. $\frac{1}{3}$

8. $\frac{2}{6}$

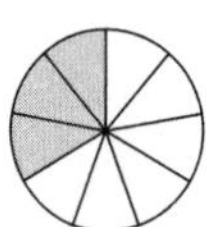

9. $\frac{3}{9}$

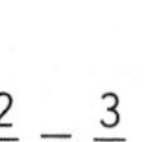

10. $\frac{1}{3} = \frac{2}{6} = \frac{3}{9}$

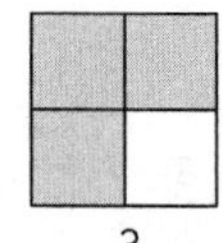

11. $\frac{3}{4}$

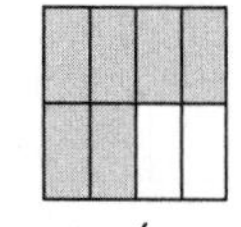

12. $\frac{6}{8}$

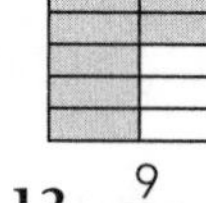

13. $\frac{9}{12}$

14. $\frac{3}{4} = \frac{6}{8} = \frac{9}{12}$

Page 29: Writing Equivalent Fractions

1. $\frac{2}{4}$
2. $\frac{2}{8}$
3. $\frac{2}{16}$
4. $\frac{4}{8}$
5. $\frac{6}{8}$
6. $\frac{2}{4}$
7. $\frac{1}{8}$
8. $\frac{6}{16}$
9. $\frac{8}{8}$
10. $\frac{3}{8}$
11. $\frac{10}{16}$
12. $\frac{1}{4}$
13. $\frac{5}{8}$
14. $\frac{2}{2}$
15. $\frac{12}{16}$
16. $\frac{1}{2}$
17. $\frac{6}{8} = \frac{12}{16}$
18. $\frac{2}{4} = \frac{8}{16}$
19. $\frac{2}{2} = \frac{8}{8}$
20. $\frac{2}{8} = \frac{1}{4}$
21. $\frac{8}{8} = \frac{4}{4}$
22. $\frac{2}{2} = \frac{1}{1}$
23. $\frac{1}{4} = \frac{4}{16}$
24. $\frac{3}{4} = \frac{6}{8}$

Page 30: More Equivalent Fractions

1. $\frac{5}{15}$
2. $\frac{9}{12}$
3. $\frac{20}{32}$
4. $\frac{8}{12}$
5. $\frac{7}{14}$
6. $\frac{12}{20}$
7. $\frac{15}{35}$
8. $\frac{10}{24}$
9. $\frac{15}{18}$
10. $\frac{49}{63}$
11. $\frac{21}{56}$
12. $\frac{28}{48}$

Page 31: Using Fractions Equal to 1

1. $\frac{2}{5} = \frac{2 \times 3}{5 \times 3} = \frac{6}{15}$ so $\frac{2}{5}$ is equal to $\frac{6}{15}$
2. $\frac{3}{4} = \frac{3 \times 4}{4 \times 4} = \frac{12}{16}$ so $\frac{3}{4}$ is equal to $\frac{12}{16}$
3. $\frac{5}{6} = \frac{5 \times 2}{6 \times 2} = \frac{10}{12}$ so $\frac{5}{6}$ is equal to $\frac{10}{12}$
4. $\frac{2}{7} = \frac{2 \times 2}{7 \times 2} = \frac{4}{14}$ so $\frac{2}{7}$ is equal to $\frac{4}{14}$
5. $\frac{1}{8} = \frac{1 \times 3}{8 \times 3} = \frac{3}{24}$ so $\frac{1}{8}$ is equal to $\frac{3}{24}$
6. $\frac{2}{9} = \frac{2 \times 3}{9 \times 3} = \frac{6}{27}$ so $\frac{2}{9}$ is equal to $\frac{6}{27}$

Page 32: Find the Numerators

1. $\frac{1}{4} = \frac{5}{20}$ because $\frac{1 \times 5}{4 \times 5} = \frac{5}{20}$
2. $\frac{3}{5} = \frac{9}{15}$ because $\frac{3 \times 3}{5 \times 3} = \frac{9}{15}$
3. $\frac{4}{7} = \frac{12}{21}$ because $\frac{4 \times 3}{7 \times 3} = \frac{12}{21}$
4. $\frac{1}{3} = \frac{5}{15}$ because $\frac{1 \times 5}{3 \times 5} = \frac{5}{15}$
5. $\frac{5}{6} = \frac{10}{12}$ because $\frac{5 \times 2}{6 \times 2} = \frac{10}{12}$
6. $\frac{2}{9} = \frac{8}{36}$ because $\frac{2 \times 4}{9 \times 4} = \frac{8}{36}$
7. $\frac{2}{3} = \frac{12}{18}$ because $\frac{2 \times 6}{3 \times 6} = \frac{12}{18}$
8. $\frac{3}{4} = \frac{36}{48}$ because $\frac{3 \times 12}{4 \times 12} = \frac{36}{48}$

Page 33: Practice

1. $\frac{1}{3} = \frac{4}{12}$
2. $\frac{3}{4} = \frac{9}{12}$
3. $\frac{1}{2} = \frac{2}{4}$
4. $\frac{2}{3} = \frac{6}{9}$
5. $\frac{1}{4} = \frac{4}{16}$
6. $\frac{1}{5} = \frac{9}{45}$
7. $\frac{1}{3} = \frac{9}{27}$
8. $\frac{1}{7} = \frac{2}{14}$
9. $\frac{2}{5} = \frac{4}{10}$
10. $\frac{1}{6} = \frac{7}{42}$
11. $\frac{7}{10} = \frac{21}{30}$
12. $\frac{5}{7} = \frac{35}{49}$
13. $\frac{7}{12} = \frac{14}{24}$
14. $\frac{11}{50} = \frac{22}{100}$
15. $\frac{9}{20} = \frac{18}{40}$
16. $\frac{7}{15} = \frac{14}{30}$
17. $\frac{11}{14} = \frac{22}{28}$
18. $\frac{7}{8} = \frac{28}{32}$
19. $\frac{5}{9} = \frac{25}{45}$
20. $\frac{1}{4} = \frac{3}{12}$

Page 34: Equivalent Fractions Review

1. $\frac{1}{3} = \frac{3}{9}$
2. $\frac{1}{3} = \frac{2}{6} = \frac{3}{9} = \frac{4}{12}$
3. $\frac{2}{5} = \frac{4}{10} = \frac{6}{15} = \frac{8}{20}$
4. $\frac{3}{5} = \frac{6}{10} = \frac{9}{15} = \frac{12}{20}$
5. a) $\frac{15}{24}$ b) $\frac{20}{24}$ c) $\frac{10}{24}$
6. a) $\frac{9}{18}$ b) $\frac{9}{18}$ c) $\frac{8}{18}$
7. a) $\frac{10}{15}$ b) $\frac{3}{15}$ c) $\frac{6}{30}$
8. a) $\frac{15}{21}$ b) $\frac{12}{16}$ c) $\frac{5}{40}$

Page 35: Less Than or Greater Than

1. a) $\frac{2}{8}$ b) $\frac{4}{8}$ c) $<$
2. a) $\frac{8}{24}$ b) $\frac{9}{24}$ c) $<$
3. a) $\frac{20}{24}$ b) $\frac{18}{24}$ c) $>$
4. a) $\frac{10}{15}$ b) $\frac{12}{15}$ c) $<$
5. a) $\frac{3}{6}$ b) $\frac{4}{6}$ c) $<$
6. a) $\frac{9}{12}$ b) $\frac{8}{12}$ c) $>$

Page 36: Compare Using the Number Line

1. $<$
2. $=$
3. $>$
4. $>$
5. $<$
6. $=$
7. $>$
8. $<$
9. $=$
10. $>$
11. $>$
12. $<$

Page 37: Connect Tags to Lines

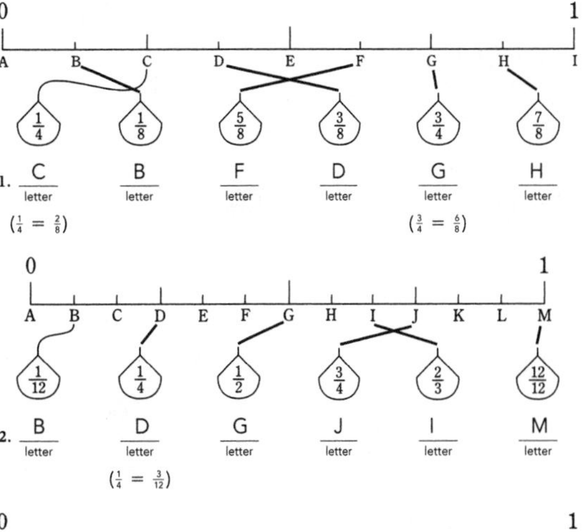

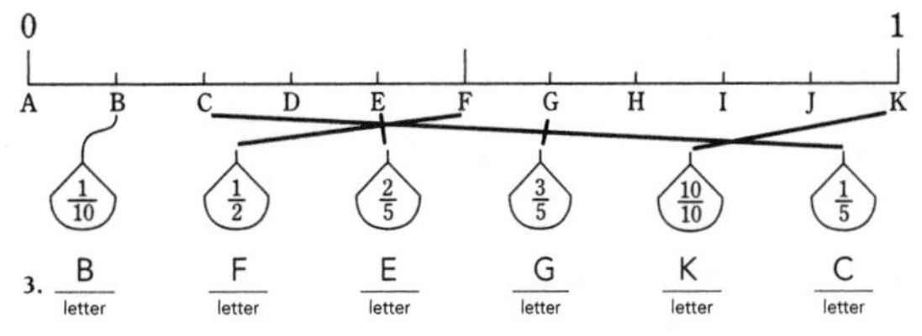

Page 38: Fractions Show Comparisons

1. $\frac{3}{5}$ **3.** $\frac{1}{3}$ **5.** $\frac{1}{2}$ **7.** $\frac{7}{10}$

2. $\frac{3}{4}$ **4.** $\frac{1}{5}$ **6.** $\frac{2}{3}$ **8.** $\frac{1}{8}$

Page 39: More Comparisons

1. $\frac{15}{30} = \frac{1}{2}$

2. $\frac{10}{20} = \frac{1}{2}$

3. $\frac{7}{10}$

4. $\frac{25}{125} = \frac{1}{5}$

5. $\frac{2}{6} = \frac{1}{3}$

6. $\frac{2}{10} = \frac{1}{5}$

7. $\frac{2}{12} = \frac{1}{6}$

8. $\frac{3}{4}$

Page 40: Comparisons Review

1. C

2. $\frac{4}{5}$ 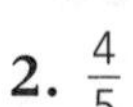 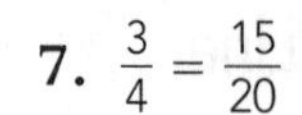

3. $\frac{3}{5}$

4. $\frac{3}{4}$

5. 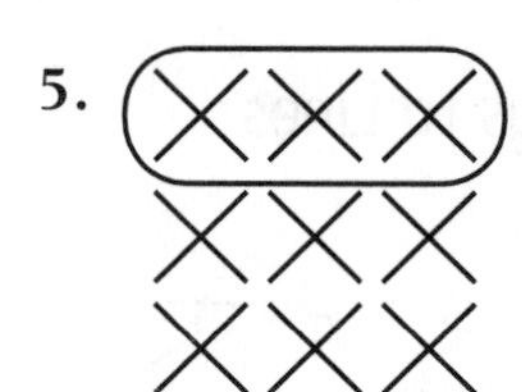

6. $1\frac{1}{4}$ 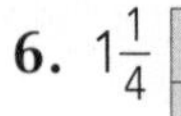 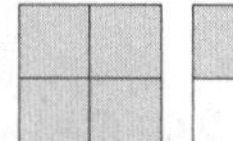

7. $\frac{3}{4} = \frac{15}{20}$

8. $\frac{5}{8} = \frac{10}{16}$

9. $\frac{1}{4}$

10. $\frac{8}{10} = \frac{4}{5}$

Page 41: Divisibility Rule for 2

The following numbers should be circled.

1. 22 **10.** 212 **15.** 1,004

5. 96 **11.** 200 **16.** 86

6. 104 **14.** 938 **18.** 100

Page 42: Divisibility Rule for 3

1. (9) **7.** 10 **13.** (12)

2. 8 **8.** (12) **14.** (9)

3. (6) **9.** (3) **15.** (9)

4. 11 **10.** (6) **16.** 7

5. 11 **11.** 10 **17.** (15)

6. 4 **12.** 13 **18.** 8

Page 43: Divisibility Rule for 5

The following numbers should be circled.

2. 55 **12.** 175

5. 15 **13.** 180

6. 50 **14.** 105

8. 35 **17.** 370

9. 80 **18.** 90

Page 44: Divisibility Rule for 10

The following numbers should be circled.

1. 30 **14.** 200

8. 70 **18.** 450

Page 45: Divisibility Practice

		Divisible by			
	Number	2	3	5	10
1.	40	✓		✓	✓
2.	144	✓	✓		
3.	94	✓			
4.	540	✓	✓	✓	✓
5.	1,000	✓		✓	✓
6.	29				
7.	45		✓	✓	
8.	85			✓	
9.	342	✓	✓		
10.	70	✓		✓	✓
11.	100	✓		✓	✓
12.	65			✓	
13.	38	✓			
14.	153		✓		
15.	384	✓	✓		
16.	89				
17.	420	✓	✓	✓	✓
18.	5,546	✓			

Page 46: Finding Factors

1. 1, 3, 5, 15
2. 4
3. 1, 2, 4, 5, 10, 20
4. 6

Page 47: Name the Factors

A. 6, 9, 18
B. 6
1. 1, 2, 3, 4, 6, 12
2. 1, 3, 9
3. 1, 2, 5, 10
4. 1, 2, 4, 7, 14, 28
5. 1, 2, 4, 8
6. 1, 2, 4
7. 1, 7
8. 1, 2, 4, 8, 16
9. 1, 2, 3, 4, 6, 8, 12, 24
10. 1, 2, 3, 5, 6, 10, 15, 30

Page 48: Greatest Common Factor

1. 1, 2, 4
2. 1, 2, 4, 8
3. 1, 2, 4
4. 4
5. 1, 2, 3, 6
6. 1, 3, 9
7. 1, 3
8. 3
9. 1, 2, 4, 8, 16
10. 1, 2, 4, 5, 10, 20
11. 1, 2, 4
12. 4

Page 49: Find the Greatest Common Factor

1. 1, 2, 5, 10
2. 1, 3, 5, 10
3. 1, 5
4. 5
5. 1, 2, 4, 8
6. 1, 2, 3, 4, 6, 12
7. 1, 2, 4
8. 4
9. 1, 2, 7, 14
10. 1, 2, 3, 4, 6, 8, 12, 24
11. 1, 2
12. 2
13. 1, 3, 7, 21
14. 1, 2, 3, 4, 6, 9, 12, 18, 36
15. 1, 3
16. 3

Page 50: Factors with Fractions

A. 1, 2, 3, 6
B. 6
1. 1, 3
2. 3
3. 1, 2, 5, 10
4. 10
5. 1, 2
6. 2
7. 1, 2, 4, 8
8. 8
9. 1, 7
10. 7

Page 51: Apply Your Skills

1. 6
2. 3
3. 2
4. 5
5. 2
6. 7
7. 8
8. 9
9. 2
10. 3
11. 2
12. 6
13. 3
14. 3
15. 3
16. 3
17. 10
18. 5
19. 5
20. 3
21. 9
22. 12
23. 11
24. 7

Page 52: Simplify Fractions

1. GCF = 7; $\frac{2}{7}$
2. GCF = 4; $\frac{1}{3}$
3. GCF = 8; $\frac{3}{4}$
4. GCF = 10; $\frac{2}{3}$
5. GCF = 11; $\frac{1}{3}$
6. GCF = 9; $\frac{2}{5}$
7. GCF = 5; $\frac{5}{6}$
8. GCF = 15; $\frac{2}{3}$
9. GCF = 4; $\frac{8}{9}$
10. GCF = 13; $\frac{2}{3}$

Page 53: Simplify

1. GCF = 6; $\frac{1}{2}$
2. GCF = 3; $\frac{1}{5}$
3. GCF = 2; $\frac{2}{3}$
4. GCF = 5; $\frac{1}{6}$
5. GCF = 2; $\frac{1}{7}$
6. GCF = 7; $\frac{2}{3}$
7. GCF = 8; $\frac{1}{3}$
8. GCF = 9; $\frac{1}{4}$
9. GCF = 2; $\frac{4}{5}$
10. GCF = 3; $\frac{2}{5}$
11. GCF = 2; $\frac{1}{2}$
12. GCF = 6; $\frac{1}{4}$
13. GCF = 3; $\frac{1}{12}$
14. GCF = 3; $\frac{4}{5}$
15. GCF = 3; $\frac{3}{4}$
16. GCF = 3; $\frac{5}{6}$
17. GCF = 10; $\frac{2}{3}$
18. GCF = 5; $\frac{5}{8}$
19. GCF = 5; $\frac{1}{10}$
20. GCF = 3; $\frac{3}{5}$
21. GCF = 9; $\frac{1}{2}$
22. GCF = 12; $\frac{1}{3}$
23. GCF = 11; $\frac{1}{2}$
24. GCF = 7; $\frac{2}{5}$

Page 54: Simplest Form

1. simplified
2. $\frac{5}{6}$
3. simplified
4. $\frac{1}{2}$
5. $\frac{2}{3}$
6. simplified
7. simplified
8. $\frac{3}{4}$
9. simplified
10. simplified
11. $\frac{1}{3}$
12. simplified
13. $\frac{4}{5}$
14. simplified
15. simplified
16. simplified
17. $\frac{3}{5}$
18. simplified
19. $\frac{3}{4}$
20. $\frac{1}{1} = 1$
21. $\frac{2}{3}$
22. $\frac{3}{4}$
23. simplified
24. $\frac{5}{6}$

Page 55: Think It Through

1. $\frac{4}{5}$
2. $\frac{1}{3}$
3. $\frac{3}{4}$
4. $\frac{3}{5}$
5. $\frac{1}{3}$
6. $\frac{2}{5}$
7. $\frac{1}{2}$
8. $\frac{1}{2}$
9. $\frac{1}{2}$
10. $\frac{2}{3}$
11. $\frac{2}{3}$
12. $\frac{2}{5}$

Page 56: Lowest Terms

1. $\frac{1}{3}$
2. $\frac{1}{4}$
3. $\frac{1}{3}$
4. $\frac{1}{4}$
5. $\frac{1}{3}$
6. $\frac{2}{3}$
7. $\frac{3}{4}$
8. $\frac{1}{2}$
9. $\frac{1}{2}$
10. $\frac{1}{3}$
11. $\frac{1}{3}$
12. $\frac{1}{3}$
13. simplified
14. $\frac{5}{6}$
15. $\frac{1}{2}$
16. $\frac{3}{4}$
17. $\frac{7}{10}$
18. $\frac{1}{2}$
19. $\frac{2}{7}$
20. $\frac{4}{9}$
21. $\frac{5}{7}$
22. simplified
23. $\frac{5}{6}$
24. $\frac{6}{7}$
25. $\frac{5}{8}$
26. $\frac{2}{3}$
27. $\frac{2}{7}$
28. simplified
29. $\frac{1}{3}$
30. $\frac{5}{7}$

Page 57: Apply Your Skills

1. $\frac{9}{10}$
2. $\frac{1}{4}$
3. $\frac{1}{2}$
4. $\frac{1}{3}$
5. $\frac{1}{8}$
6. $\frac{2}{5}$
7. $\frac{1}{13}$
8. $\frac{1}{3}$

Page 58: Divisibility and Common Factors Review

1. 63, 183, 9, 126, 249, 801
2. 155, 220, 100, 550, 345, 1,000, 335, 980
3. 1, 2, 4
4. 1, 2, 5, 10
5. 8
6. 3
7. 7
8. a) $\frac{4}{5}$
   b) $\frac{1}{3}$
9. a) $\frac{1}{3}$
   b) $\frac{1}{4}$
10. a) $\frac{2}{3}$
    b) $\frac{4}{13}$

Page 59: Changing Fractions to Decimals

1. $5\overline{)1.0}$ = .2
2. $4\overline{)3.00}$ = .75
3. $8\overline{)3.000}$ = .375
4. $8\overline{)1.000}$ = .125
5. $5\overline{)3.0}$ = .6
6. $8\overline{)7.000}$ = .875

Page 60: More Practice: Fractions to Decimals

1. 1.25
2. 1.2
3. .75
4. .5
5. .625
6. .8
7. 1.375
8. 3.75
9. .25
10. 2.4
11. .875
12. 2.2

Page 61: Working with Remainders

1. $\begin{array}{r} .66\,\frac{2}{3} \\ 3\overline{)2.00} \\ 1\,8 \\ \hline 20 \\ 18 \\ \hline 2 \end{array}$

2. $\begin{array}{r} .83\,\frac{1}{3} \\ 6\overline{)5.00} \end{array}$

3. $\begin{array}{r} .87\,\frac{1}{2} \\ 8\overline{)7.00} \end{array}$

4. $\begin{array}{r} .11\,\frac{1}{9} \\ 9\overline{)1.00} \end{array}$

5. $\begin{array}{r} .22\,\frac{2}{9} \\ 9\overline{)2.00} \end{array}$

6. $\begin{array}{r} .44\,\frac{4}{9} \\ 9\overline{)4.00} \end{array}$

Page 62: Practice Helps

1. $\begin{array}{r} .44\,\frac{4}{9} \\ 18\overline{)8.00} \\ 7\,2 \\ \hline 80 \\ 72 \\ \hline 8 \end{array}$

2. $\begin{array}{r} 2.14\,\frac{2}{7} \\ 7\overline{)15.00} \end{array}$

3. $\begin{array}{r} .85\,\frac{5}{7} \\ 7\overline{)6.00} \end{array}$

4. $\begin{array}{r} .11\,\frac{1}{9} \\ 27\overline{)3.00} \end{array}$

5. $\begin{array}{r} 1.57\,\frac{1}{7} \\ 21\overline{)33.00} \end{array}$

6. $\begin{array}{r} 1.66\,\frac{2}{3} \\ 9\overline{)15.00} \end{array}$

Page 63: Fractions and Decimals Review

1. .5
2. .75
3. 2.2
4. 3.75
5. .4
6. .625
7. 2.5
8. 3.6

Page 64: Half-Inch Measurements

1. $6\frac{1}{2}$ inches
2. $4\frac{1}{2}$ inches
3. 3 inches
4. $7\frac{1}{2}$ inches
5. $5\frac{1}{2}$ inches
6. $1\frac{1}{2}$ inches
7. $3\frac{1}{2}$ inches
8. $2\frac{1}{2}$ inches

Page 65: Quarter-Inch Measurements

1. $2\frac{1}{4}$ inches
2. $5\frac{1}{2}$ inches
3. $3\frac{3}{4}$ inches
4. $7\frac{1}{4}$ inches
5. $4\frac{1}{2}$ inches
6. $1\frac{3}{4}$ inches
7. $6\frac{1}{2}$ inches
8. $2\frac{3}{4}$ inches

Page 66: Eighth-Inch Measurements

1. $2\frac{5}{8}$ inches
2. $7\frac{3}{4}$ inches
3. $3\frac{3}{8}$ inches
4. $4\frac{3}{4}$ inches
5. $5\frac{5}{8}$ inches
6. $6\frac{1}{4}$ inches
7. $4\frac{1}{8}$ inches
8. $1\frac{7}{8}$ inches

Page 67: Draw the Measurements

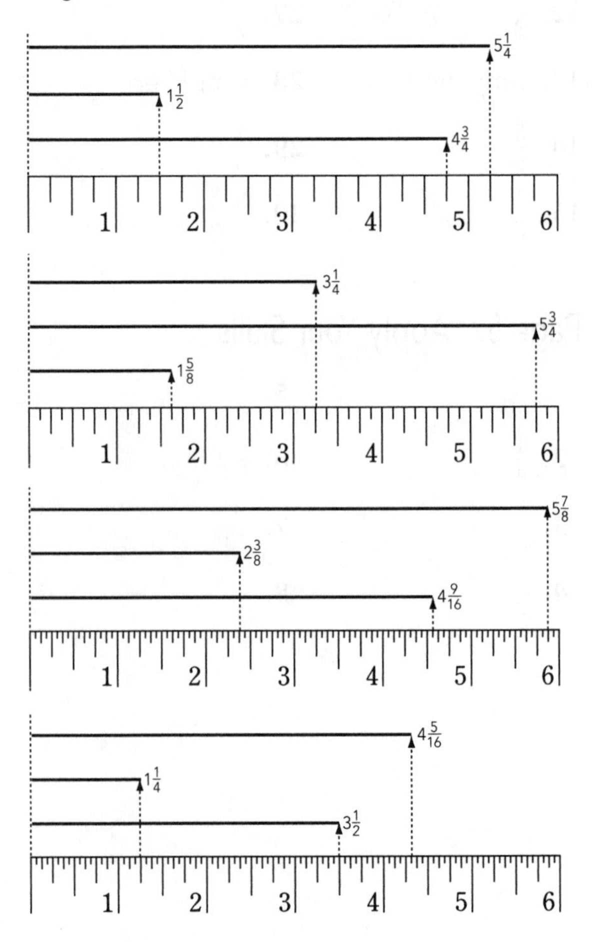

Page 68: Money as Fractions

1. $\frac{4}{10}$ or $\frac{2}{5}$
2. $\frac{10}{100}$ or $\frac{1}{10}$
3. $\frac{5}{20}$ or $\frac{1}{4}$
4. $\frac{3}{4}$
5. $\frac{5}{10}$ or $\frac{1}{2}$
6. $\frac{25}{100}$ or $\frac{1}{4}$
7. $\frac{10}{20}$ or $\frac{1}{2}$
8. $\frac{7}{20}$
9. $\frac{1}{2}$
10. $\frac{8}{10}$ or $\frac{4}{5}$

Page 69: What Coin Is Missing?

1. quarter; 25 cents
2. nickel; 5 cents
3. half-dollar; 50 cents
4. dime; 10 cents
5. quarter; 25 cents
6. nickel; 5 cents
7. nickel; 5 cents
8. half-dollar; 50 cents

Page 70: Fill Up the Gallons

A. 3 gallons
B. 20 gallons

1. $\frac{4}{5}$ 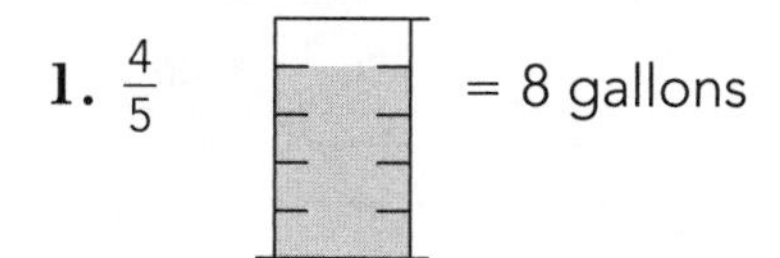 = 8 gallons
2. $\frac{1}{4}$ = 3 gallons
3. $\frac{3}{4}$ = 6 gallons
4. $\frac{3}{5}$ = 9 gallons

Page 71: Find the Measurements

1. 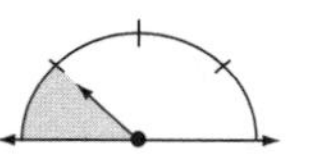
2.
3.
4.
5. 6 gallons / 0 gallons
6. 2 gallons
7. 4 gallons
8. 6 gallons
9. 3 gallons
10. 8 gallons / 0 gallons
11. 2 gallons
12. 4 gallons
13. 6 gallons
14. 8 gallons

Page 72: Measurement Review

1. $3\frac{7}{8}$
2. $2\frac{1}{4}$
3. $4\frac{3}{4}$
4. $5\frac{9}{16}$

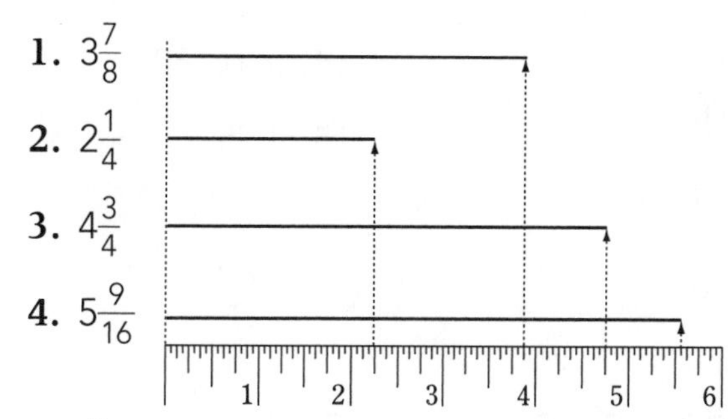

5. $\frac{9}{20}$

6. Empty 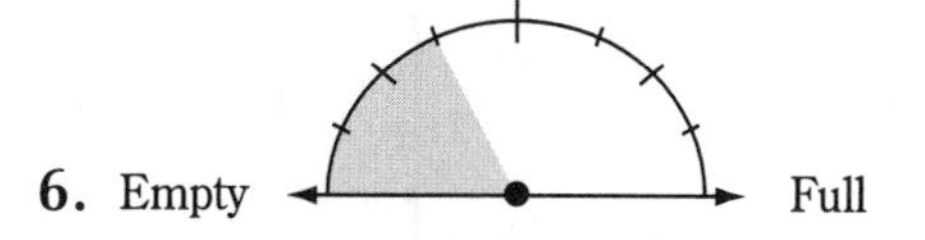 Full

7. dime; 10 cents   nickel; 5 cents

8. 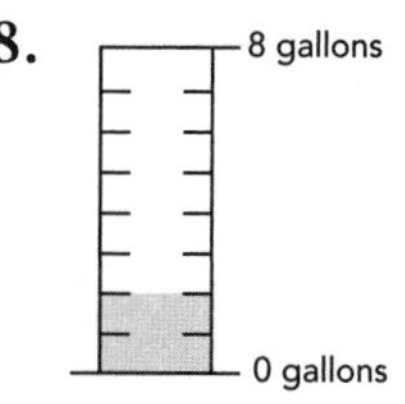

Page 73: Real-Life Applications

1. 30
2. 10
3. 20
4. 75
5. 6
6. 3
7. 4
8. 18
9. 50
10. 25
11. 75
12. 10

Page 74: Life-Skills Math Review

1. 10
2. 16
3. a) 3 oranges
   b) 1 gallon
   c) 4 cups
4. $\frac{50}{100}$ or $\frac{1}{2}$
5. $\frac{2}{3}$
6. $\frac{1}{7}$
7. $\frac{4}{5}$
8. $\frac{1}{2}$
9. $\frac{5}{6}$
10. $\frac{2}{5}$
11. $\frac{3}{5}$
12. $\frac{8}{15}$
13. .875
14. .8 or .80
15. 1.25

Page 75: Cumulative Review

1. 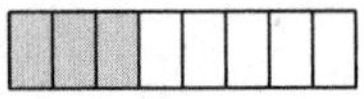
2. <
3. 3
4. 6
5. $\frac{7}{10}$
6. .875
7. 1.75
8. $\frac{3}{5}$
9. 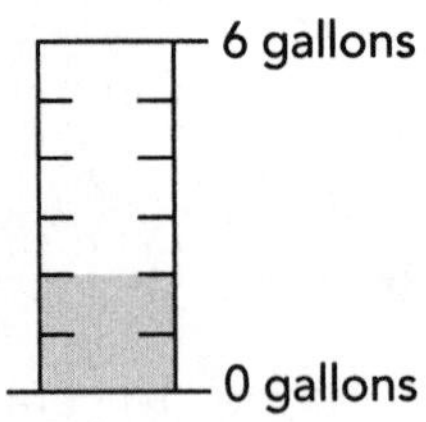

10. 15

Page 76: Posttest

1. $\frac{1}{9}$
2. $\frac{10}{30}$
3. $\frac{2}{3}$
4. $\frac{7}{4}$ and $\frac{13}{4}$
5. .625
6. $\frac{7}{8}$
7. $\frac{5}{12}$
8. $\frac{1}{5}$
9. $\frac{20}{24}$
10. 2.6
11. $\frac{2}{5}$
12. $\frac{25}{40}$
13. $\frac{1}{9}$
14. 1.35
15. $\frac{44}{48}$
16. $\frac{13}{20}$
17. $\frac{5}{12}$
18. $\frac{4}{5}$
19. $\frac{5}{9}$
20. $.41\frac{2}{3}$

Posttest Evaluation Chart

If a student misses one or more problems in a skill area, assign a review of the practice pages for that skill.

Skill Area	Posttest Problem #	Skill Section	Review Page
Identifying Fractions	4, 13	7–19	20
Comparing Fractions	1, 6	21–25	26
Equivalent Fractions	2, 9, 12, 15	27–33	34
Comparisons	8, 11, 16, 18	35–39	40
Simplifying Fractions	3, 7, 17, 19	41–57	58
Fractions and Decimals	5, 10, 14, 20	59–62	63
Life-Skills Math	All	64–73	72, 74

FRACTIONS
Addition & Subtraction

The activities in this section will encourage students to discover for themselves how addition and subtraction of fractions work. Some students need many experiences with diagrams, shading pictorial models, etc., while others already understand the concepts and are capable of moving more quickly.

Student Glossary

Acquainting students with definitions of key math terms and life-skills concepts will enhance their mastery of the materials. Below are words defined in the student text. A glossary is provided at the end of the student text and on page 243 of the *Teacher's Resource Guide and Answer Key.*

appointment
comparison
denominator
factor
fuel gauge
gallon
greatest common factor
hour
improper fraction
least common multiple
like fractions
line segment
lowest common denominator
mile
mixed number
multiple
number relation symbol
numerator
overtime
portion
pound
rename
simplify (reduce)
symbol
whole number

STUDENT PAGE
7
Addition

Using a Number Line

Introduce how a ruler shows fractions. Then show students how to count segments. This allows them to discover that when adding fractions, **denominators** are the same. Only the **numerators** are added.

EXAMPLE $\frac{3}{4} + \frac{3}{4} = \frac{6}{4}$ or $1\frac{1}{2}$

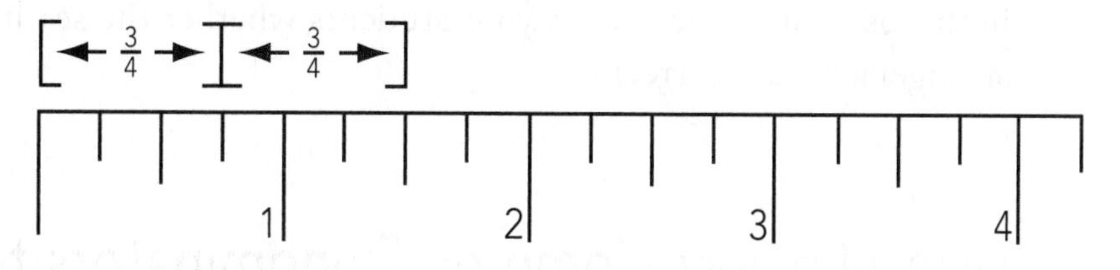

STUDENT PAGE
17
Lowest Common Denominator

Flash-Card Drill

On a series of flash cards, write two **unlike fractions.** Have students name the **lowest common denominator.**

$\frac{1}{3}$ $\frac{1}{2}$	$\frac{2}{5}$ $\frac{1}{10}$	$\frac{3}{4}$ $\frac{1}{3}$
(Answer: 6)	(Answer: 10)	(Answer: 12)

STUDENT PAGE
18
Lowest Common Denominator

Different Denominators

Write fractions with different denominators on the board. Students should identify the largest fraction with a square and the smallest with a circle. To do this, have them rename each set of fractions with like denominators.

EXAMPLES

A original fractions: $\frac{1}{2}$ $\frac{1}{6}$ $\frac{2}{3}$ $\frac{7}{8}$

renamed fractions: $\frac{12}{24}$ $\frac{4}{24}$ (circled) $\frac{16}{24}$ $\frac{21}{24}$ (boxed)

B $\frac{3}{4}$ $\frac{1}{2}$ $\frac{1}{4}$ $\frac{2}{3}$

C $\frac{1}{16}$ $\frac{3}{8}$ $\frac{1}{2}$ $\frac{3}{4}$

STUDENT PAGE

19

Lowest Common Denominator

Comparing and Ordering Information

Place fractions smaller than 1 on note cards for each student in the class. Randomly pass out one card per student. Students in each row should get together and compare fractions. The student with the smallest fraction sits in the first seat. The student with the next smallest fraction sits in the second seat. Have students continue until the student with the largest fraction sits in the last seat. Discuss with your students whether the seating arrangements are correct.

STUDENT PAGE

20

Lowest Common Denominator

Using Lowest Common Denominators to Add Money

You can use students' understanding of money to help them see why they need to find a lowest common denominator to add fractions.

1 Fractions of a dollar can be written and simplified when compared to 100 pennies. Demonstrate and discuss the relationships below by using real or play money.

$1 = \frac{1}{100}$ of a dollar

$5 = \frac{5}{100} = \frac{1}{20}$ of a dollar

$10 = \frac{10}{100} = \frac{1}{10}$ of a dollar

$25 = \frac{25}{100} = \frac{1}{4}$ of a dollar

$50 = \frac{50}{100} = \frac{1}{2}$ of a dollar

2 Review how to find equivalent fractions. Students will discover that in order to add fractions they must have **like denominators**. Have students explain what must be done with the denominators before they can add two fractions.

Use one example with money to demonstrate:

- changing to equivalent fractions
- adding the numerators

EXAMPLES

1 quarter and 1 half-dollar $= \frac{1}{4} + \frac{1}{2} = \frac{25}{100} + \frac{50}{100} = \frac{75}{100} = \frac{3}{4}$

1 dime and 1 quarter =

1 quarter and 1 nickel =

STUDENT PAGE
21
Lowest Common Denominator

Estimate Your Answers

In daily life we often need to estimate fractions. Have students round one of the mixed numbers in a pair and add. Remind students that a fraction of $\frac{1}{2}$ or more rounds the mixed number to the next highest whole number.

EXAMPLES

$7\frac{2}{3}$ rounds up to 8

$+\ 3\frac{1}{2} \longrightarrow +\ 3\frac{1}{2}$

$11\frac{1}{2}$

$5\frac{5}{6} \longrightarrow 5\frac{5}{6}$

$+\ 2\frac{7}{8}$ rounds up to $+\ 3$

$8\frac{5}{6}$

STUDENT PAGE
22
Lowest Common Denominator

Addition Review

Have students complete the table. You may add more fractions and mixed numbers along the top and side.

+	$\frac{1}{4}$	$\frac{3}{4}$	$9\frac{1}{2}$
$\frac{3}{4}$	1		
$\frac{2}{3}$			
$2\frac{1}{5}$			

STUDENT PAGE
27
Subtraction

Number Lines

The number line is a good way to model subtraction. Lined notebook paper turned on its side can help students make equal divisions for their drawings.

1 Give students a fraction problem and a number line. Let them use the number line to find the answer.

EXAMPLE $\frac{4}{5} - \frac{2}{5} = \frac{2}{5}$

? $\frac{2}{5}$

$\frac{4}{5}$

0 $\frac{1}{5}$ $\frac{2}{5}$ $\frac{3}{5}$ $\frac{4}{5}$ 1

2 Have students make their own number lines to model each problem.

EXAMPLES

$\frac{5}{7} - \frac{3}{7} =$

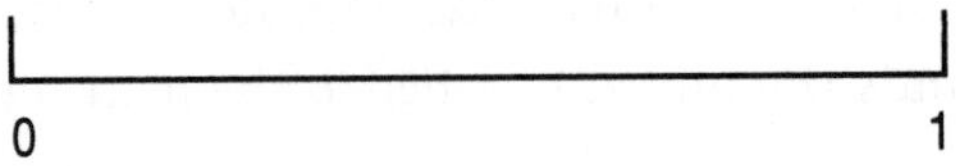

$\frac{4}{6} - \frac{3}{6} =$

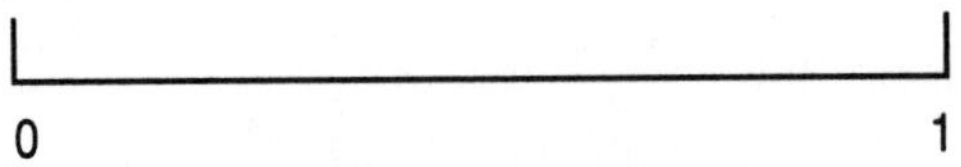

3 Let students find the difference between two points on a number line. Tell them to count the spaces between the points.

EXAMPLE

space between the points

$1\frac{2}{4} - \frac{3}{4} = \frac{3}{4}$

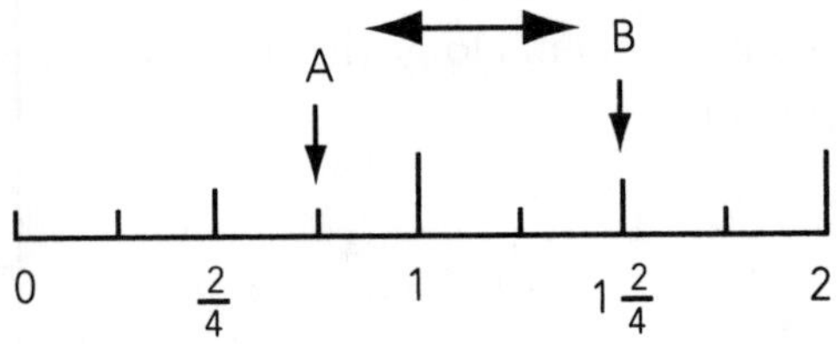

STUDENT PAGE **28** Subtraction

Using Regions

Using graph paper, have students subtract pairs of fractions with the same denominators.

EXAMPLES

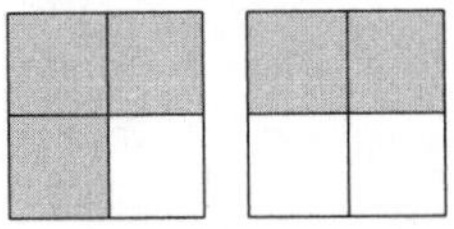

A $\frac{3}{4} - \frac{2}{4} = \frac{1}{4}$

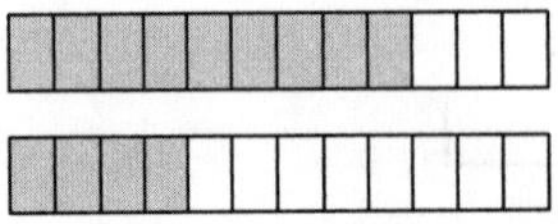

B $\frac{9}{12} - \frac{4}{12} = \frac{5}{12}$

STUDENT PAGE
46
Subtraction

Find the Difference

Write a set of fractions on flash cards or the board. Ask students a series of questions that requires them to manipulate the numbers. Remind them to start by giving the fractions the same denominator.

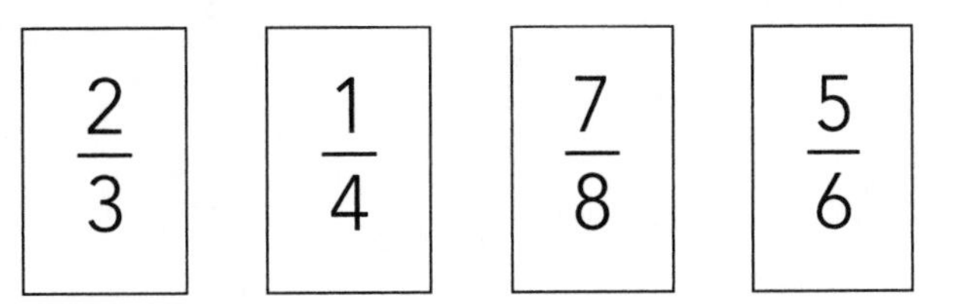

SAMPLE QUESTIONS

What two fractions will result in the largest difference?

What two fractions will result in the smallest difference?

What two fractions will result in the largest sum?

What two fractions will result in the smallest sum?

Will the two fractions with the smallest difference also be the two fractions with the largest sum?

Alternate Activity

Have students write six different subtraction problems from four given fractions. Then have them give the problems to a classmate to solve.

Page 4: Pretest

1. $\frac{7}{8}$
2. $9\frac{2}{5}$
3. $14\frac{1}{2}$
4. 10
5. $\frac{11}{12}$
6. $7\frac{13}{16}$
7. $16\frac{7}{12}$
8. $\frac{1}{3}$
9. $3\frac{1}{4}$
10. $4\frac{11}{8}$
11. $3\frac{1}{3}$
12. $2\frac{5}{6}$
13. $5\frac{1}{3}$
14. $\frac{8}{15}$
15. $1\frac{1}{4}$ hours
16. $2\frac{1}{4}$ hours
17. $1\frac{1}{2}$ yards
18. $\frac{1}{2}$ of the pizza
19. $25\frac{5}{8}$ pounds
20. $4\frac{13}{16}$ miles

Pretest Evaluation Chart

If a student misses any problems in a skill area, assign the practice pages for that skill. However, students may need to complete all practice pages to reinforce areas of weakness.

Skill Area	Pretest Problem #	Skill Section	Review Page
Addition	1, 2, 3, 4, 5, 6, 7	7–30	18, 31, 57, 58
Subtraction	8, 9, 10, 11, 12, 13, 14	39–55	56, 57, 58
Addition Problem Solving	15, 18, 19	32–37	38, 66, 67
Subtraction Problem Solving	16, 17, 20	59–65	66, 67
Life-Skills Math	All	68–73	74

Page 7: Compare and Add

1. $\frac{3}{9}$
2. $\frac{4}{9}$
3. $\frac{7}{9}$

Page 8: Add the Fractions

1. $\frac{1}{4} + \frac{1}{4} = \frac{2}{4}$ or $\frac{1}{2}$
2. $\frac{3}{4} + \frac{1}{4} = \frac{4}{4}$ or 1
3. $\frac{2}{8} + \frac{3}{8} = \frac{5}{8}$
4. $\frac{3}{8} + \frac{4}{8} = \frac{7}{8}$
5. $\frac{6}{16} + \frac{6}{16} = \frac{12}{16}$ or $\frac{3}{4}$
6. $\frac{6}{16} + \frac{5}{16} = \frac{11}{16}$

Page 9: Like Denominators

1. $\frac{6}{7}$
2. $\frac{11}{14}$
3. $\frac{5}{6}$
4. $\frac{9}{11}$
5. $\frac{14}{17}$
6. $\frac{13}{15}$
7. $\frac{11}{15}$
8. $\frac{11}{18}$
9. $\frac{3}{13}$
10. $\frac{9}{11}$

Page 10: Add and Simplify

1. $\frac{2}{4} = \frac{1}{2}$
2. $\frac{5}{10} = \frac{1}{2}$
3. $\frac{9}{12} = \frac{3}{4}$
4. $\frac{10}{15} = \frac{2}{3}$
5. $\frac{15}{18} = \frac{5}{6}$
6. $\frac{15}{21} = \frac{5}{7}$
7. $\frac{5}{15} = \frac{1}{3}$
8. $\frac{6}{9} = \frac{2}{3}$
9. $\frac{15}{20} = \frac{3}{4}$

Page 11: Adding Mixed Numbers

1. $12\frac{3}{9} = 12\frac{1}{3}$
2. $8\frac{6}{8} = 8\frac{3}{4}$
3. $11\frac{5}{8}$
4. $9\frac{6}{12} = 9\frac{1}{2}$
5. $11\frac{9}{10}$
6. $7\frac{2}{4} = 7\frac{1}{2}$
7. $13\frac{7}{9}$
8. $10\frac{3}{6} = 10\frac{1}{2}$
9. $21\frac{4}{10} = 21\frac{2}{5}$
10. $6\frac{5}{7}$
11. $25\frac{12}{16} = 25\frac{3}{4}$
12. $11\frac{8}{15}$

Page 12: Fractions Greater Than 1

1. $\frac{7}{4} = 1\frac{3}{4}$
2. $\frac{11}{5} = 2\frac{1}{5}$
3. $\frac{11}{3} = 3\frac{2}{3}$
4. $\frac{19}{8} = 2\frac{3}{8}$

Page 13: Change to Mixed Numbers

1. $2\frac{2}{4} = 2\frac{1}{2}$
2. $1\frac{1}{6}$
3. $3\frac{4}{6} = 3\frac{2}{3}$
4. $1\frac{4}{5}$
5. $4\frac{1}{5}$
6. $6\frac{2}{6} = 6\frac{1}{3}$
7. $2\frac{1}{3}$
8. $2\frac{5}{10} = 2\frac{1}{2}$
9. $2\frac{1}{5}$
10. $6\frac{3}{9} = 6\frac{1}{3}$
11. 6
12. $7\frac{1}{2}$

Page 14: Practice Simplifying

1. $2\frac{4}{3} = 2 + \frac{4}{3}$
   $= 2 + 1\frac{1}{3}$
   $= 3\frac{1}{3}$
2. $5\frac{6}{5} = 5 + \frac{6}{5}$
   $= 5 + 1\frac{1}{5}$
   $= 6\frac{1}{5}$
3. $6\frac{13}{8} = 6 + \frac{13}{8}$
   $= 6 + 1\frac{5}{8}$
   $= 7\frac{5}{8}$
4. $11\frac{13}{6} = 11 + \frac{13}{6}$
   $= 11 + 2\frac{1}{6}$
   $= 13\frac{1}{6}$
5. $8\frac{19}{9} = 8 + \frac{19}{9}$
   $= 8 + 2\frac{1}{9}$
   $= 10\frac{1}{9}$
6. $7\frac{23}{8} = 7 + \frac{23}{8}$
   $= 7 + 2\frac{7}{8}$
   $= 9\frac{7}{8}$

Page 15: Simplify

1. $3\frac{5}{6}$
2. $2\frac{5}{12}$
3. $10\frac{3}{4}$
4. 5
5. $4\frac{1}{4}$
6. 8
7. $11\frac{1}{4}$
8. $8\frac{5}{7}$
9. $8\frac{1}{3}$
10. $9\frac{1}{3}$
11. $7\frac{2}{3}$
12. $14\frac{1}{5}$
13. 17
14. $22\frac{2}{3}$
15. $2\frac{4}{5}$
16. $21\frac{1}{2}$
17. $24\frac{2}{3}$
18. $6\frac{1}{2}$

Page 16: Simplify Your Answers

1. $\frac{7}{5} = 1\frac{2}{5}$
2. $\frac{8}{7} = 1\frac{1}{7}$
3. $\frac{5}{4} = 1\frac{1}{4}$
4. $\frac{7}{6} = 1\frac{1}{6}$
5. $\frac{12}{8} = 1\frac{4}{8} = 1\frac{1}{2}$
6. $\frac{16}{12} = 1\frac{4}{12} = 1\frac{1}{3}$
7. $\frac{12}{9} = 1\frac{3}{9} = 1\frac{1}{3}$
8. $\frac{12}{10} = 1\frac{2}{10} = 1\frac{1}{5}$

Page 17: Simplify When Necessary

1. $\frac{10}{8} = 1\frac{2}{8} = 1\frac{1}{4}$
2. $6\frac{8}{9}$
3. $\frac{9}{6} = 1\frac{3}{6} = 1\frac{1}{2}$
4. $6\frac{4}{16} = 6\frac{1}{4}$
5. $8\frac{10}{14} = 8\frac{5}{7}$
6. $15\frac{1}{4}$
7. $13\frac{7}{11}$
8. $4\frac{10}{15} = 4\frac{2}{3}$
9. $\frac{15}{13} = 1\frac{2}{13}$
10. $15\frac{4}{4} = 16$
11. $\frac{11}{7} = 1\frac{4}{7}$
12. $8\frac{16}{10} = 9\frac{6}{10} = 9\frac{3}{5}$

Page 18: Addition Review

1. $\frac{11}{13}$
2. $\frac{5}{15} = \frac{1}{3}$
3. $6\frac{2}{3}$
4. $16\frac{13}{12} = 17\frac{1}{12}$
5. $\frac{14}{3} = 4\frac{2}{3}$
6. $\frac{10}{8} = 1\frac{1}{4}$
7. 11
8. $\frac{8}{10} = \frac{4}{5}$
9. $3\frac{1}{4}$
10. $9\frac{1}{6}$

Page 19: Multiples

1. 8, 16, 24, 32, 40, 48, 56, 64, 72, 80
2. 8, 16, 24, 32, 40, 48, 56, 64, 72, 80
3. 2, 4, 6, 8, 10, 12, 14, 16, 18, 20
4. 3, 6, 9, 12, 15, 18, 21, 24, 27, 30
5. 4, 8, 12, 16, 20, 24, 28, 32, 36, 40
6. 6, 12, 18, 24, 30, 36, 42, 48, 54, 60
7. 9, 18, 27, 36, 45, 54, 63, 72, 81, 90
8. 12, 24, 36, 48, 60, 72, 84, 96, 108, 120

Page 20: Least Common Multiple

1. 4, 8, 12, 16, 20, 24, 28, 32, 36, 40
2. 6, 12, 18, 24, 30, 36, 42, 48, 54, 60
3. 12, 24, 36
4. 12
5. 3, 6, 9, 12, 15, 18, 21, 24, 27, 30
6. 5, 10, 15, 20, 25, 30, 35, 40, 45, 50
7. 15, 30
8. 15
9. 8, 16, 24, 32, 40, 48, 56, 64, 72, 80
10. 10, 20, 30, 40, 50, 60, 70, 80, 90, 100
11. 40, 80
12. 40

Page 21: Lowest Common Denominator

1. 4, 8, 12, 16, 20, 24, 28, 32
2. 5, 10, 15, 20, 25, 30, 35, 40
3. 20
4. 6, 12, 18, 24, 30, 36, 42, 48
5. 8, 16, 24, 32, 40, 48, 56, 64
6. 24
7. 7, 14, 21, 28, 35, 42, 49, 56
8. 4, 8, 12, 16, 20, 24, 28, 32
9. 28

Page 22: Find the Lowest Common Denominator (LCD)

1. 4, 8, 12
2. 12, 24, 36
3. 12
4. 6, 12, 18
5. 3, 6, 9
6. 6
7. 5, 10, 15
8. 10, 20, 30
9. 10
10. 8, 16, 24, 32
11. 2, 4, 6, 8
12. 8
13. 6, 12, 18
14. 4, 8, 12
15. 12
16. 9, 18, 27
17. 3, 6, 9
18. 9

Page 23: Choose the Best Method

1. Method 2
LCD is 24

2. Method 2
LCD is 15

3. Method 1
LCD is 12

4. Method 2
LCD is 18

5. Method 1
LCD is 9

6. Method 1
LCD is 22

Page 24: Fraction Readiness

1. LCD = 6
$\frac{2}{3} = \frac{4}{6}$
$\frac{1}{2} = \frac{3}{6}$

2. LCD = 12
$\frac{2}{3} = \frac{8}{12}$
$\frac{3}{4} = \frac{9}{12}$

3. LCD = 10
$\frac{3}{5} = \frac{6}{10}$
$\frac{7}{10} = \frac{7}{10}$

4. LCD = 18
$\frac{8}{9} = \frac{16}{18}$
$\frac{5}{6} = \frac{15}{18}$

5. LCD = 12
$\frac{1}{4} = \frac{3}{12}$
$\frac{5}{6} = \frac{10}{12}$

6. LCD = 24
$\frac{5}{12} = \frac{10}{24}$
$\frac{3}{8} = \frac{9}{24}$

7. LCD = 4
$\frac{1}{2} = \frac{2}{4}$
$\frac{1}{4} = \frac{1}{4}$

8. LCD = 24
$\frac{1}{8} = \frac{3}{24}$
$\frac{5}{6} = \frac{20}{24}$

Page 25: Compare Unlike Fractions

1. $\frac{1}{3} = \frac{4}{12}$

2. $\frac{1}{4} = \frac{3}{12}$

3. $\frac{1}{3} > \frac{1}{4}$

4. $\frac{1}{2} = \frac{3}{6}$

5. $\frac{5}{6} = \frac{5}{6}$

6. $\frac{1}{2} < \frac{5}{6}$

7. $\frac{2}{3} = \frac{10}{15}$

8. $\frac{4}{5} = \frac{12}{15}$

9. $\frac{2}{3} < \frac{4}{5}$

10. $\frac{5}{6} = \frac{5}{6}$

11. $\frac{2}{3} = \frac{4}{6}$

12. $\frac{5}{6} > \frac{2}{3}$

13. $\frac{3}{8} = \frac{3}{8}$

14. $\frac{3}{4} = \frac{6}{8}$

15. $\frac{3}{8} < \frac{3}{4}$

16. $\frac{1}{3} = \frac{8}{24}$

17. $\frac{3}{8} = \frac{9}{24}$

18. $\frac{1}{3} < \frac{3}{8}$

Page 26: Compare Using Symbols

1. $<$

2. $>$

3. $>$

4. $<$

5. $<$

6. $<$

7. $>$

8. $<$

9. $>$

10. $>$

11. $<$

12. $>$

Page 27: Adding with Unlike Denominators

1. $\frac{4}{6} + \frac{1}{6} = \frac{5}{6}$

2. $\frac{6}{10} + \frac{2}{10} = \frac{8}{10} = \frac{4}{5}$

3. $\frac{1}{12} + \frac{9}{12} = \frac{10}{12} = \frac{5}{6}$

4. $1\frac{7}{12}$

5. $\frac{7}{9}$

6. $\frac{4}{5}$

7. $1\frac{5}{21}$

8. $1\frac{1}{4}$

9. $1\frac{3}{20}$

Page 28: Adding Mixed Numbers with Unlike Denominators

1. $7\frac{3}{15} + 3\frac{10}{15} = 10\frac{13}{15}$
2. $2\frac{3}{6} + 5\frac{4}{6} = 7\frac{7}{6} = 8\frac{1}{6}$
3. $3\frac{11}{12} + 4\frac{9}{12} = 7\frac{20}{12} = 8\frac{8}{12} = 8\frac{2}{3}$
4. $6\frac{1}{2}$
5. $16\frac{1}{4}$
6. $9\frac{13}{20}$
7. $4\frac{13}{14}$
8. $20\frac{1}{5}$
9. $28\frac{2}{3}$

Page 29: Mixed Practice

1. $\frac{13}{24}$
2. $9\frac{1}{2}$
3. $12\frac{7}{24}$
4. $17\frac{1}{5}$
5. $28\frac{4}{5}$
6. $33\frac{1}{2}$
7. $1\frac{7}{12}$
8. $37\frac{5}{6}$
9. $26\frac{2}{3}$
10. $30\frac{11}{18}$
11. $8\frac{2}{3}$
12. $42\frac{1}{3}$

Page 30: Adding Three Numbers

1. $3\frac{2}{8} + 1\frac{1}{8} + 5\frac{4}{8} = 9\frac{7}{8}$
2. $1\frac{4}{5}$
3. $13\frac{7}{24}$
4. $1\frac{1}{4}$
5. $18\frac{1}{5}$
6. $1\frac{5}{18}$
7. $19\frac{5}{8}$
8. $14\frac{3}{4}$
9. $10\frac{11}{20}$

Page 31: Lowest Common Denominator Review

1. 4, 8, 12, 16, 20, 24
2. 6, 12, 18, 24, 30, 36
3. 12
4. 4
5. 18
6. $\frac{1}{4} < \frac{2}{5}$
7. $\frac{8}{10} > \frac{3}{5}$
8. $4\frac{13}{14}$
9. $27\frac{1}{2}$
10. $18\frac{1}{4}$

Page 32: Does the Answer Make Sense?

1. $4\frac{5}{8} + 7\frac{1}{3} = 11\frac{23}{24}$; $11\frac{23}{24}$
2. $6\frac{1}{2} + 3\frac{1}{4} = 9\frac{3}{4}$; $9\frac{3}{4}$
3. $1\frac{1}{2} + \frac{3}{4} = 2\frac{1}{4}$; $2\frac{1}{4}$
4. $\frac{3}{4} + 12\frac{1}{2} = 18\frac{1}{4}$; $18\frac{1}{4}$
5. $1\frac{5}{8} + 2\frac{3}{4} = 3\frac{11}{8} = 4\frac{3}{8}$; $4\frac{3}{8}$
6. $8\frac{1}{2} + 2\frac{1}{2} = 11$; 11

Page 33: Number Sentences

1. $\frac{3}{4} + \frac{1}{2} = 1\frac{1}{4}$; $1\frac{1}{4}$
2. $\frac{1}{3} + \frac{1}{4} = \frac{7}{12}$; $\frac{7}{12}$
3. $1\frac{3}{4} + 2\frac{1}{2} = 4\frac{1}{4}$; $4\frac{1}{4}$
4. $1\frac{1}{2} + \frac{1}{4} = 1\frac{3}{4}$; $1\frac{3}{4}$
5. $\frac{3}{5} + \frac{1}{5} = \frac{4}{5}$; $\frac{4}{5}$
6. $\frac{1}{3} + \frac{1}{5} = \frac{8}{15}$; $\frac{8}{15}$

Page 34: Picture Problems

1. 6 pounds
2. $8\frac{1}{4}$ miles
3. $11\frac{3}{4}$ miles
4. $1\frac{1}{12}$ cups
5. $14\frac{1}{2}$
6. $5\frac{3}{4}$ feet

Page 35: Using Symbols

1. $4\frac{1}{3} > 3\frac{3}{4}$
2. $4\frac{1}{3} + 3\frac{3}{4} = 8\frac{1}{12}$ miles
3. $3\frac{3}{4} < 4\frac{1}{3}$
4. $4\frac{1}{3} \neq 3\frac{3}{4}$
5. $7\frac{5}{8} > 2\frac{1}{4}$
6. $7\frac{5}{8} + 2\frac{1}{4} = 9\frac{7}{8}$ yards
7. $2\frac{1}{4} < 7\frac{5}{8}$
8. $7\frac{5}{8} \neq 2\frac{1}{4}$

Page 36: Write a Question

1. What do the pieces of wood measure altogether?
2. a) How much did the chocolate and the hard candy weigh?
   b) How much did all the chocolate, peanuts, and hard candy weigh altogether?
   c) How much did the chocolate and the peanuts weigh together?
3. a) How much time does the entire trip take?
   b) How much time is spent on the plane and the bus?
4. How tall are Carol and Marc altogether?
5. a) How far did they drive in all?
   b) How far did Matilda and Cloe drive?
6. How long were both lines combined?

Page 37: Addition Word Problems

1. $3\frac{1}{2} + 1\frac{2}{3} = 5\frac{1}{6}$
2. $\frac{3}{4} + \frac{1}{2} = 1\frac{1}{4}$
3. $2\frac{1}{4} + \frac{3}{4} = 3$
4. $3\frac{1}{2} + 2\frac{1}{8} = 5\frac{5}{8}$
5. $5\frac{1}{4} + 1\frac{2}{3} = 6\frac{11}{12}$
6. $23\frac{3}{8} + 15\frac{3}{4} = 39\frac{1}{8}$

Page 38: Addition Problem-Solving Review

1. $14\frac{3}{4} + 3\frac{3}{4} = 18\frac{1}{2}$
2. $34\frac{1}{2} + 38\frac{3}{4} = 73\frac{1}{4}$
3. $20\frac{1}{4} + 25\frac{1}{2} = 45\frac{3}{4}$
4. $3\frac{1}{2} + 5\frac{3}{4} = 9\frac{1}{4}$
5. $\frac{12}{12} + \frac{4}{12} = 1\frac{1}{3}$
6. $1\frac{1}{2} + \frac{1}{4} + \frac{3}{4} + 1\frac{1}{3} = 3\frac{5}{6}$

Page 39: How Much Is Left?

1. $\frac{1}{5}$
2. $\frac{3}{10}$
3. $\frac{1}{6}$
4. subtract

Page 40: Take Away the Fractions

1. $\frac{3}{4} - \frac{2}{4} = \frac{1}{4}$
2. $\frac{7}{8} - \frac{4}{8} = \frac{3}{8}$
3. $\frac{11}{16} - \frac{6}{16} = \frac{5}{16}$

Page 41: Subtracting with Like Denominators

1. $\frac{3}{7}$
2. $\frac{3}{5}$
3. $\frac{1}{4}$
4. $\frac{2}{5}$
5. $\frac{2}{3}$
6. $\frac{3}{7}$
7. $\frac{3}{5}$
8. $\frac{1}{2}$
9. $\frac{1}{6}$
10. $\frac{6}{11}$
11. $\frac{4}{5}$
12. $\frac{5}{7}$

Page 42: Subtracting Mixed Numbers

1. $2\frac{3}{5}$
2. $5\frac{1}{3}$
3. $7\frac{2}{3}$
4. $3\frac{5}{8}$
5. $4\frac{4}{8} = 4\frac{1}{2}$
6. $3\frac{3}{5}$
7. $5\frac{1}{2}$
8. $9\frac{3}{20}$
9. $6\frac{6}{10} = 6\frac{3}{5}$
10. $8\frac{7}{8}$
11. $6\frac{2}{11}$
12. $10\frac{1}{5}$

Page 43: Subtracting Fractions with Unlike Denominators

1. $\frac{1}{15}$
2. $\frac{9}{12} - \frac{8}{12} = \frac{1}{12}$
3. $\frac{5}{6} - \frac{2}{6} = \frac{3}{6} = \frac{1}{2}$
4. $\frac{4}{12} = \frac{1}{3}$
5. $\frac{2}{9}$
6. $\frac{13}{18}$
7. $\frac{1}{8}$
8. $\frac{5}{10} = \frac{1}{2}$
9. $\frac{9}{12} = \frac{3}{4}$

Page 44: Subtracting Mixed Numbers with Unlike Denominators

1. $7\frac{9}{10} - 3\frac{5}{10} = 4\frac{4}{10} = 4\frac{2}{5}$
2. $8\frac{6}{10} - \frac{1}{10} = 7\frac{5}{10} = 7\frac{1}{2}$
3. $6\frac{8}{12} - 2\frac{3}{12} = 4\frac{5}{12}$
4. $4\frac{5}{6}$
5. $2\frac{1}{15}$
6. $3\frac{5}{10} = 3\frac{1}{2}$
7. $2\frac{4}{12} = 2\frac{1}{3}$
8. $4\frac{2}{6} = 4\frac{1}{3}$
9. $6\frac{3}{5}$

Page 45: Subtraction Practice

1. $6\frac{5}{6}$
2. $5\frac{4}{6} = 5\frac{2}{3}$
3. $4\frac{3}{24} = 4\frac{1}{8}$
4. $\frac{13}{18}$
5. $\frac{4}{10} = \frac{2}{5}$
6. $9\frac{5}{15} = 9\frac{1}{3}$
7. $\frac{3}{12} = \frac{1}{4}$
8. $3\frac{1}{24}$
9. $3\frac{2}{12} = 3\frac{1}{6}$
10. $\frac{5}{10} = \frac{1}{2}$
11. $2\frac{7}{8}$
12. $5\frac{4}{30} = 5\frac{2}{15}$

Page 46: Renaming Whole Numbers

1. $3\frac{4}{4}$
2. $2\frac{8}{8}$
3. $4\frac{2}{2}$
4. $2\frac{5}{5}$

Page 47: Mastering the Skill

1. $7\frac{3}{3}$
2. $5\frac{8}{8}$
3. $1\frac{9}{9}$
4. $2\frac{6}{6}$
5. $3\frac{7}{7}$
6. $2\frac{4}{4}$
7. $10\frac{5}{5}$
8. $6\frac{9}{9}$
9. $\frac{10}{10}$
10. $5\frac{14}{14}$
11. $12\frac{15}{15}$
12. $1\frac{3}{3}$
13. $4\frac{2}{2}$
14. $16\frac{4}{4}$
15. $3\frac{9}{9}$
16. $8\frac{7}{7}$
17. $7\frac{12}{12}$
18. $5\frac{20}{20}$

Page 48: Rename and Subtract

1. $5\frac{3}{3} - 2\frac{2}{3} = 3\frac{1}{3}$
2. $3\frac{2}{2} - 3\frac{1}{2} = \frac{1}{2}$
3. $1\frac{3}{5}$
4. $1\frac{1}{8}$
5. $2\frac{3}{10}$
6. $7\frac{1}{12}$
7. $6\frac{1}{4}$
8. $5\frac{2}{7}$
9. $3\frac{1}{8}$
10. $3\frac{1}{5}$
11. $\frac{7}{8}$
12. $1\frac{3}{4}$

Page 49: Rename Mixed Numbers

1. $2\frac{5}{4}$
2. $1\frac{4}{3}$
3. $3\frac{3}{2}$
4. $2\frac{11}{6}$

Page 50: Mixed Number Readiness

1. $4\frac{11}{8}$

   $4 + \frac{8}{8} + \frac{3}{8}$

   $4\frac{11}{8}$
2. $6\frac{8}{5}$

   $6 + 1 + \frac{3}{5}$

   $6 + \frac{5}{5} + \frac{3}{5}$

   $6\frac{8}{5}$
3. $3\frac{3}{2}$

   $3 + 1 + \frac{1}{2}$

   $3 + \frac{2}{2} + \frac{1}{2}$

   $3\frac{3}{2}$
4. $7\frac{5}{3}$

   $7 + 1 + \frac{2}{3}$

   $7 + \frac{3}{3} + \frac{2}{3}$

   $7\frac{5}{3}$

Page 51: Renaming Shortcut

1. $5\frac{15}{8}$
2. $3\frac{14}{9}$
3. $12\frac{4}{3}$
4. $8\frac{9}{5}$
5. $6\frac{11}{6}$
6. $1\frac{18}{14}$
7. $6\frac{27}{20}$
8. $4\frac{25}{18}$
9. $12\frac{17}{10}$
10. $18\frac{9}{7}$

Page 52: Rename to Subtract

1. $6\frac{11}{8} - 4\frac{5}{8} = 2\frac{6}{8} = 2\frac{3}{4}$
2. $5\frac{7}{5} - \frac{4}{5} = 5\frac{3}{5}$
3. $4\frac{2}{6} = 4\frac{1}{3}$
4. $7\frac{5}{6}$
5. $1\frac{4}{7}$
6. $3\frac{7}{10}$
7. $1\frac{2}{4} = 1\frac{1}{2}$
8. $6\frac{4}{5}$
9. $3\frac{6}{9} = 3\frac{2}{3}$

Page 53: Subtract Unlike Mixed Numbers

1. $\begin{array}{r} 7\frac{1}{4} = 7\frac{2}{8} = 6\frac{10}{8} \\ -\ 5\frac{3}{8} = 5\frac{3}{8} = 5\frac{3}{8} \\ \hline 1\frac{7}{8} \end{array}$

2. $2\frac{3}{6} = 2\frac{1}{2}$

3. $2\frac{9}{10}$

4. $1\frac{10}{12} = 1\frac{5}{6}$

5. $3\frac{5}{9}$

6. $4\frac{5}{18}$

7. $6\frac{8}{10} = 6\frac{4}{5}$

8. $7\frac{5}{6}$

Page 54: Decide to Rename

1. $4\frac{7}{10}$

2. $4\frac{7}{8}$

3. $2\frac{4}{6} = 2\frac{2}{3}$

4. $8\frac{9}{12} = 8\frac{5}{12}$

5. $4\frac{3}{6} = 4\frac{1}{2}$

6. $6\frac{23}{40}$

7. $3\frac{3}{12} = 3\frac{1}{4}$

8. $7\frac{14}{30} = 7\frac{7}{15}$

Page 55: Mixed Practice

1. $\frac{5}{8}$

2. $4\frac{3}{6} = 4\frac{1}{2}$

3. $6\frac{9}{16}$

4. $6\frac{3}{10}$

5. $6\frac{5}{12}$

6. $2\frac{8}{10} = 2\frac{4}{5}$

7. $8\frac{7}{12}$

8. $4\frac{2}{3}$

9. $10\frac{1}{6}$

10. $5\frac{13}{15}$

11. $\frac{5}{12}$

12. $\frac{1}{3}$

Page 56: Subtraction Review

1. $\frac{2}{13}$

2. $1\frac{1}{11}$

3. $\frac{7}{12}$

4. $2\frac{1}{15}$

5. $6\frac{1}{6}$

6. $6\frac{4}{4}$

7. $1\frac{2}{5}$

8. $1\frac{3}{2}$

9. $1\frac{2}{3}$

10. $4\frac{5}{18}$

11. $1\frac{3}{8}$

12. $2\frac{9}{11}$

Page 57: Mixed Addition and Subtraction

1. $8\frac{19}{14} = 9\frac{5}{14}$

2. $\frac{24}{20} = 1\frac{4}{20} = 1\frac{1}{5}$

3. $8\frac{2}{6} = 8\frac{1}{3}$

4. $12\frac{12}{10} = 13\frac{2}{10} = 13\frac{1}{5}$

5. $7\frac{13}{20}$

6. $8\frac{28}{30} = 8\frac{14}{15}$

7. $9\frac{3}{6} = 9\frac{1}{2}$

8. $\frac{7}{12}$

9. $4\frac{3}{6} = 4\frac{1}{2}$

10. $16\frac{20}{12} = 17\frac{8}{12} = 17\frac{2}{3}$

11. $8\frac{5}{15} = 8\frac{1}{3}$

12. $32\frac{29}{18} = 33\frac{11}{18}$

Page 58: Putting It All Together

1. $\frac{3}{8} < \frac{5}{8}$

2. $1\frac{1}{2} = 1\frac{1}{2}$

3. $9\frac{1}{8} > 8\frac{1}{6}$

4. $1\frac{1}{3} > 1\frac{1}{6}$

5. $11\frac{1}{2} > 11\frac{1}{20}$

6. $\frac{13}{15} < 1\frac{3}{4}$

Page 59: Subtraction Word Problems

1. $5\frac{1}{2} - 1\frac{1}{4} = 4\frac{1}{4}$; $4\frac{1}{4}$

2. $\frac{3}{4} - \frac{1}{4} = \frac{1}{2}$; $\frac{1}{2}$

3. $1\frac{1}{4} - \frac{1}{2} = \frac{3}{4}$; $\frac{3}{4}$

4. $4\frac{1}{3} - 1\frac{1}{2} = 2\frac{5}{6}$; $2\frac{5}{6}$

5. $3\frac{3}{4} - 1\frac{1}{2} = 2\frac{1}{4}$; $2\frac{1}{4}$

6. $8 - \frac{5}{8} = 7\frac{3}{8}$; $7\frac{3}{8}$

Page 60: Writing Number Sentences

1. $15\frac{2}{3} - 4\frac{1}{4} = 11\frac{5}{12}$; $11\frac{5}{12}$
2. $7\frac{1}{2} - 4\frac{3}{4} = 2\frac{3}{4}$; $2\frac{3}{4}$
3. $3\frac{1}{2} - 1\frac{1}{4} = 2\frac{1}{4}$; $2\frac{1}{4}$
4. $20\frac{1}{2} - 3\frac{1}{3} = 17\frac{1}{6}$; $17\frac{1}{6}$
5. $1\frac{1}{3} - \frac{3}{4} = \frac{7}{12}$; $\frac{7}{12}$
6. $8\frac{1}{2} - 6\frac{3}{8} = 2\frac{1}{8}$; $2\frac{1}{8}$

Page 61: Use Drawings to Solve Problems

1. $7\frac{5}{6}$ miles
2. $21\frac{3}{4}$ miles

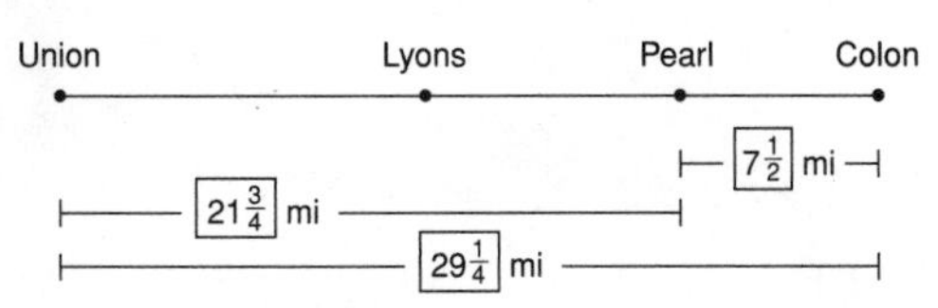

3. $3\frac{1}{4}$ miles

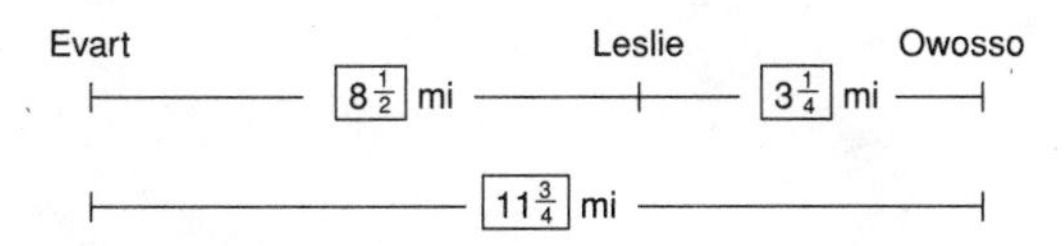

Page 62: Picture Problems

1. $8\frac{3}{4} - 1\frac{1}{2} = 7\frac{1}{4}$
2. $2 - 1\frac{1}{4} = \frac{3}{4}$
3. $\frac{3}{8} + \frac{1}{4} = \frac{5}{8}$
4. $3\frac{1}{3} + \frac{3}{4} =$ ~~$3\frac{1}{12}$~~ $4\frac{1}{12}$ (handwritten correction)
5. $16\frac{1}{4} - 10\frac{3}{4} = 5\frac{2}{4} = 5\frac{1}{2}$
6. $13\frac{1}{2} + 2\frac{3}{4} = 15\frac{5}{4} = 16\frac{1}{4}$

Page 63: Add or Subtract

1. $1\frac{1}{2} + 2\frac{1}{3} = 3\frac{5}{6}$; $3\frac{5}{6}$
2. $12 - 2\frac{3}{8} = 9\frac{5}{8}$; $9\frac{5}{8}$
3. $12 - 7\frac{3}{4} = 4\frac{1}{4}$; $4\frac{1}{4}$
4. $6\frac{1}{2} + \frac{13}{16} = 6\frac{21}{16}$; $7\frac{5}{16}$

Page 64: Using Symbols

1. $\frac{3}{4} + 1\frac{1}{4} = 2$
2. $2 - 1\frac{1}{4} = \frac{3}{4}$
3. $2 > \frac{3}{4}$
4. $\frac{3}{4} < 2$
5. $\frac{3}{4} \neq 2$
6. $2\frac{1}{2} + 2\frac{3}{4} = 5\frac{1}{4}$
7. $5\frac{1}{4} - 2\frac{3}{4} = 2\frac{1}{2}$
8. $5\frac{1}{4} > 2\frac{1}{2}$
9. $2\frac{1}{2} < 5\frac{1}{4}$
10. $5\frac{1}{4} \neq 2\frac{1}{2}$

Page 65: Write a Question

Questions will vary but should ask for the following items:

1. a) sum of $6\frac{3}{8}$ and $3\frac{1}{4}$
   b) difference of $6\frac{3}{8}$ and $3\frac{1}{4}$
2. a) difference of $3\frac{1}{2}$ and $2\frac{3}{4}$
   b) difference of $3\frac{1}{2}$ and $1\frac{1}{4}$
3. a) difference of $3\frac{1}{2}$ and $1\frac{1}{3}$
   b) sum of $3\frac{1}{2}$ and $1\frac{1}{3}$
4. a) difference between 5 and $\frac{7}{8}$
   b) difference between 7 and $\frac{1}{2}$
5. a) difference of $2\frac{1}{4}$ and $1\frac{1}{2}$
   b) sum of $2\frac{1}{4}$ and $1\frac{1}{2}$
6. a) difference of 4 and $1\frac{1}{4}$
   b) difference of 4 and $2\frac{1}{2}$

Page 66: Decide to Add or Subtract

Answers will vary.

Page 67: Mixed Problem-Solving Review

1. $\frac{7}{8} - \frac{1}{4} = \frac{5}{8}; \frac{5}{8}$
2. $4\frac{5}{8} + 7\frac{9}{16} = 11\frac{19}{16} = 12\frac{3}{16}; 12\frac{3}{16}$
3. $2\frac{3}{8} - 1\frac{1}{3} = 1\frac{1}{24}; 1\frac{1}{24}$
4. $4 - 2\frac{1}{8} = 1\frac{7}{8}; 1\frac{7}{8}$
5. $1\frac{1}{2} + \frac{1}{4} = 1\frac{3}{4}; 1\frac{3}{4}$
6. $1\frac{1}{3} + \frac{3}{5} = 1\frac{14}{15}; 1\frac{14}{15}$

Page 68: Weighing Food

A.

1. a) 7
   b) ~~Z~~ Y
   c) $5\frac{1}{2}$ pounds
   d) $4\frac{3}{4}$ pounds
2. $3\frac{1}{4} - \frac{3}{8} = 2\frac{7}{8}$
3. $6 - 3\frac{3}{4} = 2\frac{1}{4}$

Page 69: Reading a Measuring Cup

A.

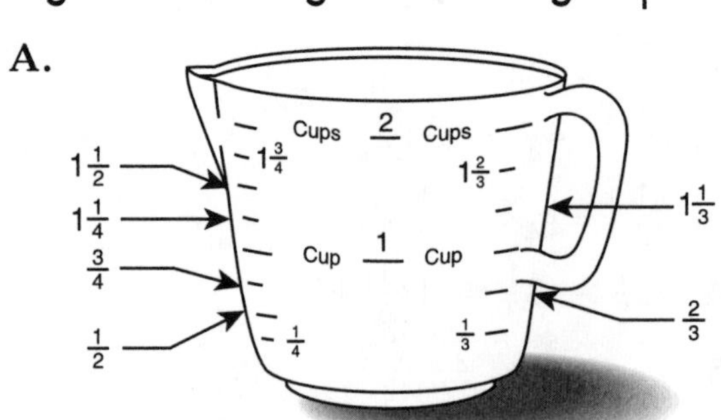

1. 
   a) $\frac{3}{4}$ cup

   b) $\frac{1}{3}$ cup

   c) $\frac{2}{3}$ cup

   d) $\frac{1}{2}$ cup
2. $1\frac{1}{2}$ times
3. a) $\frac{1}{2}$ cup
   b) $\frac{1}{4}$ cup
   c) $1\frac{3}{4}$ cup
   d) $\frac{3}{4}$ cup

Page 70: Math for the Carpenter

1. 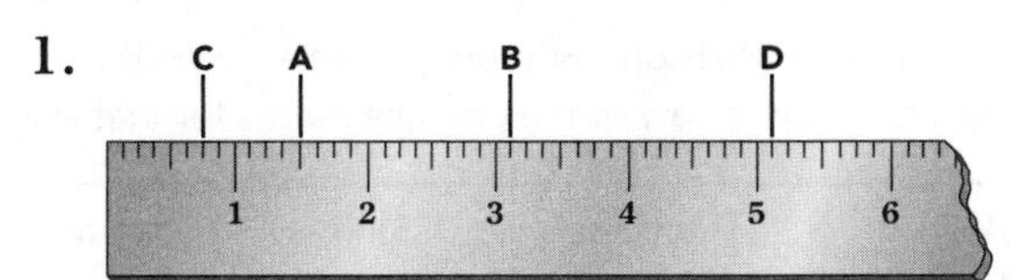

2. a) 1 inch

   b) $1\frac{7}{8}$ inches

3. C
4. nail should be 2 inches long

Page 71: Fractions of an Hour

1. $1\frac{1}{2}$
2. 1:30
3. a) 15
   b) 1:45
4. 12:15
5. a) 30
   b) 2:00
6. $5 - 1\frac{1}{2} = 3\frac{1}{2}$ hours

Page 72: Using Road Signs

1. Cedar Lake $6\frac{3}{4}$ miles
2. Fulton $1\frac{3}{4}$ miles

   Scotts 10 miles

3. Big Rapids $4\frac{1}{4}$ miles
4. Dublin $8\frac{5}{8}$ miles

   Alma $8\frac{1}{8}$ miles

Page 73: Apply Your Skills

1. $7\frac{3}{4}$ pounds
2. $2\frac{1}{12}$ cups
3. 12:30
4. 4:00
5. $3\frac{3}{4}$ hours
6. $80\frac{1}{6}$ miles
7. $101\frac{1}{4}$ miles
8. 1:45

Page 74: Life-Skills Math Review

1. $1\frac{1}{4}$ pounds
2. 

3. $1\frac{3}{4}$ cups
4. $6\frac{3}{4}$
5. a) 45 minutes
   b) 12:30
6. $4\frac{3}{4}$ hours
7. $11\frac{3}{4}$ miles
8. $6\frac{1}{2}$ miles

Page 75: Cumulative Math Review

1. 1
2. $2\frac{7}{8}$
3. $18\frac{13}{14}$
4. $20\frac{1}{5}$
5. $15\frac{1}{8}$
6. $\frac{11}{17}$
7. $\frac{1}{3}$
8. $\frac{3}{4}$
9. $7\frac{2}{3}$
10. $1\frac{1}{3}$

Page 76: Posttest

1. $\frac{5}{6}$
2. $1\frac{3}{4}$ hours
3. $10\frac{13}{15}$
4. $3\frac{3}{4}$ hours
5. $3\frac{1}{6}$
6. $7\frac{1}{4}$
7. $\frac{1}{8}$
8. $31\frac{7}{10}$ miles
9. $29\frac{11}{16}$ inches
10. $6\frac{5}{12}$
11. 14
12. $8\frac{19}{12}$
13. $11\frac{5}{8}$
14. $2\frac{5}{6}$
15. $1\frac{5}{24}$
16. $\$\frac{3}{4}$ million
17. $7\frac{1}{4}$ hours
18. $2\frac{7}{8}$
19. $7\frac{15}{16}$ pounds
20. $13\frac{1}{8}$ miles

Posttest Evaluation Chart

If a student misses one or more problems in a skill area, assign a review of the practice pages for that skill.

Skill Area	Posttest Problem #	Skill Section	Review Page
Addition	1, 3, 6, 11, 13, 15	7–30	18, 31, 57, 58
Subtraction	5, 7, 10, 12, 14, 18	39–55	56, 57, 58
Addition Problem Solving	4, 9, 17, 19	32–37	38, 66, 67
Subtraction Problem Solving	2, 8, 16, 20	59–65	66, 67
Life-Skills Math	All	68–73	74

FRACTIONS
Multiplication & Division

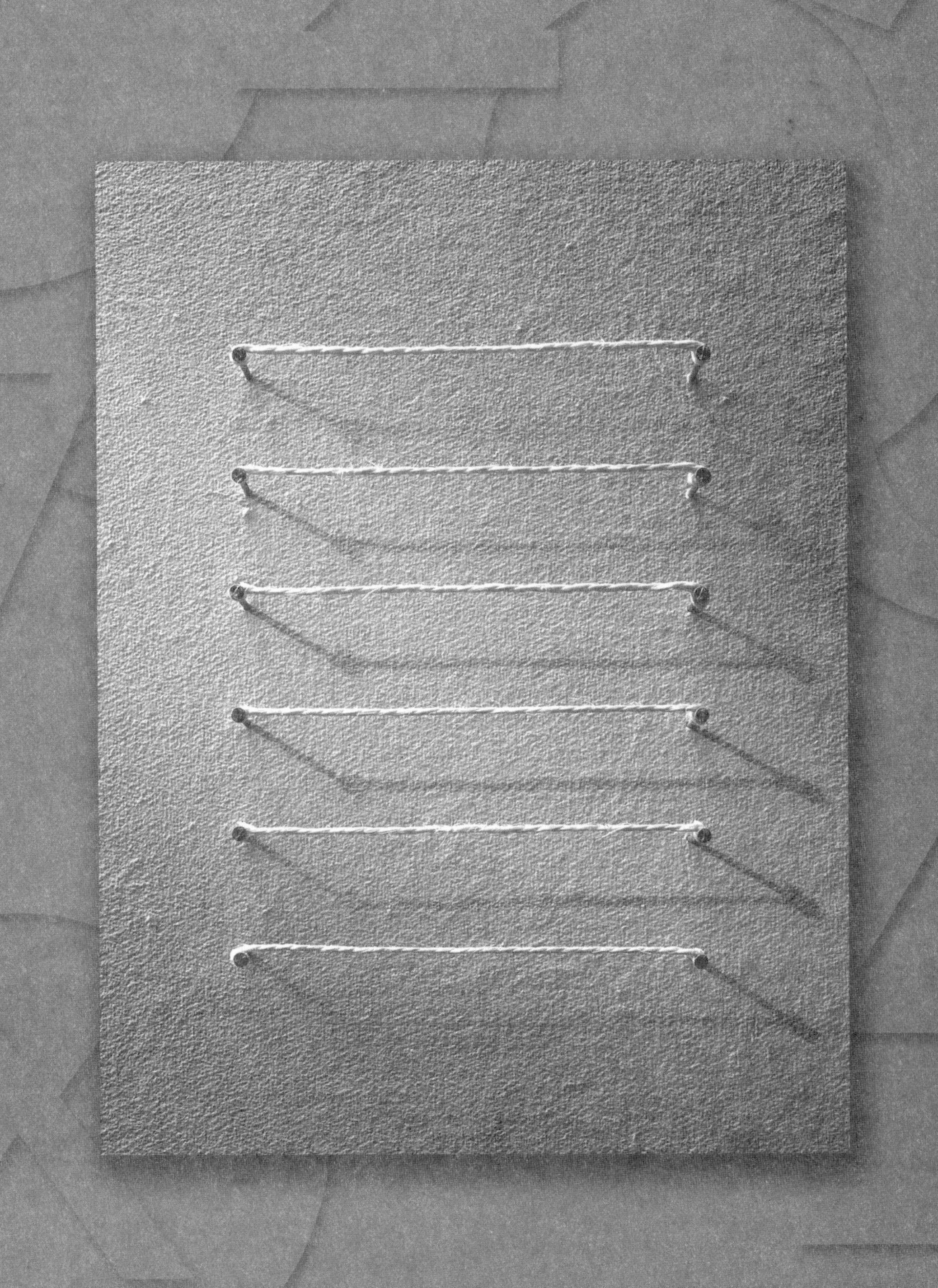

The activities in this section will encourage students to discover for themselves how multiplication and division of fractions work. Some students need many experiences with diagrams, shading pictorial models, etc., while others already understand the concepts and are capable of moving more quickly.

CONTENTS OF STUDENT TEXT	PAGE NUMBERS

Student Glossary

Acquainting students with definitions of key math terms and life-skills concepts will enhance their mastery of the materials. Below are words defined in the student text. A glossary is provided at the end of the student text and on page 243 of the *Teacher's Resource Guide and Answer Key.*

average
batch
centimeter
denominator
discount
earnings
greatest common factor
improper fraction
income
ingredients
line segment
mixed number
number relation symbol
numerator
oz (ounce)
product
rate
reciprocal
recreation
rename
savings account
shift
simplify (reduce)
survey
symbol
tbsp (tablespoon)
tsp (teaspoon)
whole number

STUDENT PAGE
8
Understanding Multiplication of Fractions

Finding a Fraction of a Number

Tell students:

$\frac{1}{2}$ times anything is the same as dividing it by 2.

$\frac{1}{3}$ times anything is the same as dividing it by 3.

Remind students that the word *of* means *to multiply*. For example, the math problem $\frac{1}{2}$ of 6 = 3 can also be written $\frac{1}{2} \times 6 = 3$. Have students practice finding the fraction of a number. Begin by rounding dollar amounts to whole numbers to make it easy to divide.

EXAMPLES

$\frac{1}{2} \times \$9.75$ Think $\frac{1}{2} \times \$10 = \$$______

$\frac{1}{3} \times \$20.87$ Think $\frac{1}{3} \times \$21 = \$$______

Explain that this technique can be used when shopping to determine the sale price of an item.

STUDENT PAGE
9
Understanding Multiplication of Fractions

Number Lines

To help students get a concrete sense of multiplication by fractions, have them use a combination of pictorial models and number sentences.

1 Draw a pictorial model based on a number sentence.

EXAMPLE Draw $5 \times \frac{1}{4} = \frac{5}{4}$

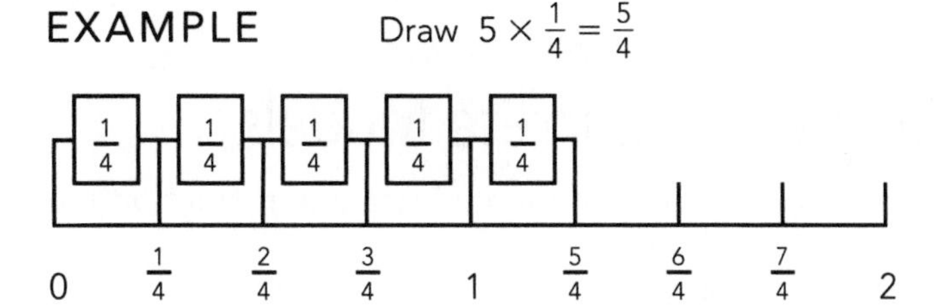

2 Write a number sentence based on a pictorial model.

EXAMPLE Write a number sentence based on this picture.

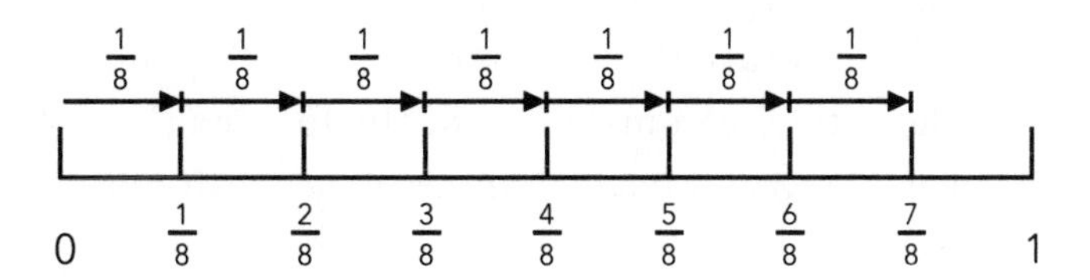

Answer: $7 \times \frac{1}{8} = \frac{7}{8}$

3 Have students draw a picture and demonstrate fraction relationships. Ask questions based on the model.

EXAMPLE What is $\frac{1}{3}$ of 6? Answer: $\frac{1}{3} \times 6 = 2$

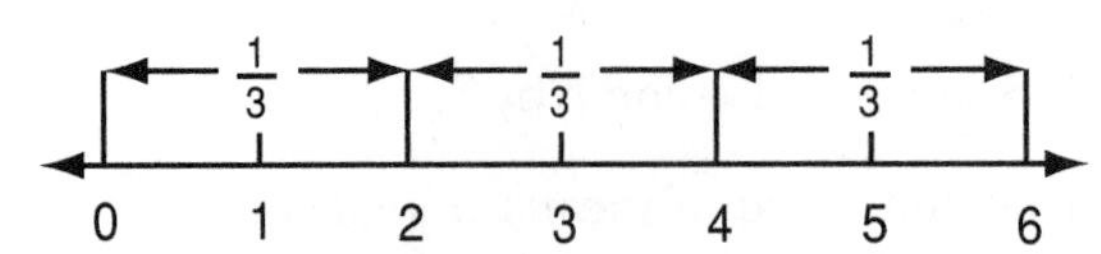

Alternate Activity

Provide number lines of 0–8 with marks for whole and half numbers or have students make their own using a ruler. Ask them to fold the strip in half. Ask students:

What is $\frac{1}{2}$ of 8? _______

Next, cut off 1 inch so that a 7-inch strip remains. Ask them to fold this strip in half. Then ask:

What is $\frac{1}{2}$ of 7? _______

Continue with this procedure until only 1 inch remains. Ask:

What is $\frac{1}{2}$ of 1? _______

Is multiplying by $\frac{1}{2}$ the same as dividing by 2? _______

STUDENT PAGE
13
Understanding Multiplication of Fractions

Working with Concrete Models

Use manipulatives to help students learn about fraction multiplication.

EXAMPLE $4 \times \frac{3}{4} = 3$

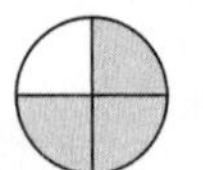 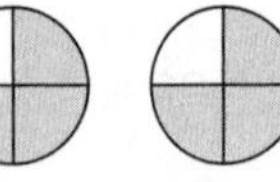 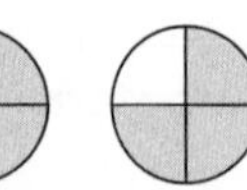 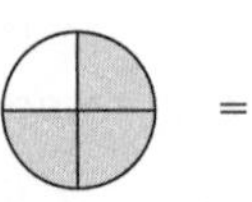 = 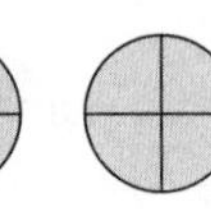 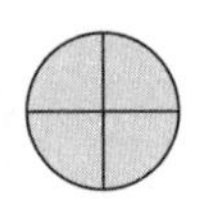

Use 4 paper plates to represent the whole number 4. Have students cut each plate into fourths and save three-fourths from each plate. Ask students how many whole plates they can make out of the fourths they saved and if there are any left over. They will have 3 whole plates with nothing left over.

STUDENT PAGE
14
Understanding Multiplication of Fractions

Using Regions to Show Multiplication by Fractions

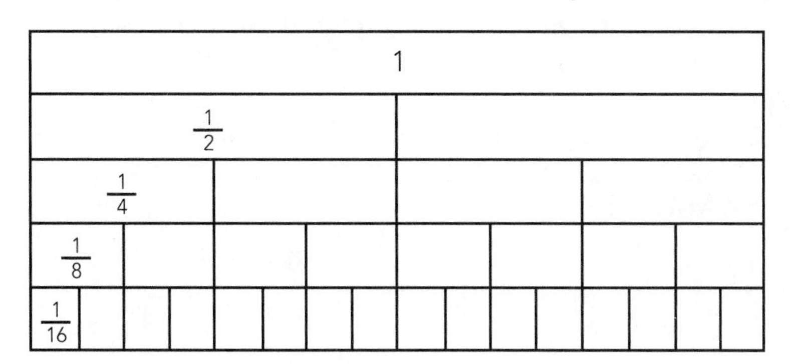

Show illustrations of a region that is separated into halves, fourths, eighths, and sixteenths. Have students shade or color one-half of 1, one-half of $\frac{1}{2}$, etc. Summarize the exercise by writing numerical expressions such as:

$\frac{1}{2}$ of $1 = \frac{1}{2}$ $\frac{1}{2}$ of $\frac{1}{4} = \frac{1}{8}$

$\frac{1}{2}$ of $\frac{1}{2} = \frac{1}{4}$ $\frac{1}{2}$ of $\frac{1}{8} = \frac{1}{16}$

STUDENT PAGE
14
Understanding Multiplication of Fractions

Pictorial Models of Multiplying Fractions

Another valuable model for teaching multiplication of fractions is a rectangular unit subdivided into smaller parts. When students begin to work with these problems, they will discover that the **product** will be less than both factors.

EXAMPLE $\frac{1}{3}$ of $\frac{1}{4} = \frac{1}{12}$

Step 1: Show $\frac{1}{4}$

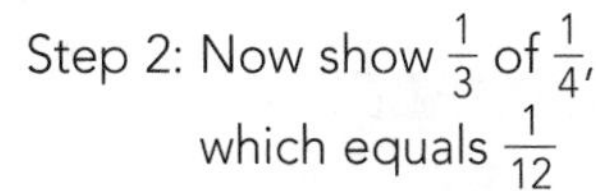

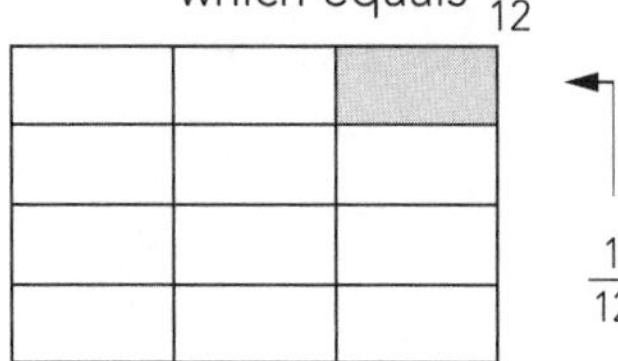

Students enjoy this activity once they get the idea. Work several problems with the class and then let them try some on their own. This is another way to visualize multiplication with fractions.

FRACTIONS: Multiplication and Division

STUDENT PAGE
17
Multiplication

Seeing Multiplication Factors

Have students work backward from the answer to see the factors in a multiplication problem.

Use two of the three fractions to make a true number sentence.

EXAMPLE 1

$\frac{1}{5}$ $\frac{1}{2}$ $\frac{1}{4}$

A $\frac{\square}{\square} \times \frac{\square}{\square} = \frac{1}{20}$ **B** $\frac{\square}{\square} \times \frac{\square}{\square} = \frac{1}{8}$ **C** $\frac{\square}{\square} \times \frac{\square}{\square} = \frac{1}{10}$

EXAMPLE 2

$\frac{2}{3}$ $\frac{4}{7}$ $\frac{2}{5}$

A $\frac{\square}{\square} \times \frac{\square}{\square} = \frac{4}{15}$ **B** $\frac{\square}{\square} \times \frac{\square}{\square} = \frac{8}{21}$ **C** $\frac{\square}{\square} \times \frac{\square}{\square} = \frac{8}{35}$

STUDENT PAGE
27
Multiplication

Estimate for a Reasonable Product

Students should learn to round each mixed number to the nearest whole number and multiply the rounded whole numbers. Remind students that mixed numbers with fractions of $\frac{1}{2}$ or more should be rounded up to the next highest whole number.

EXAMPLES

$2\frac{1}{4} \times 5\frac{3}{5}$ rounds to $2 \times 6 = 12$

$5\frac{1}{3} \times 2\frac{1}{2}$ rounds to $5 \times 3 =$ ______

$3\frac{5}{9} \times 2\frac{1}{8}$ rounds to ______ × ______ = ______

$7\frac{3}{4} \times 3\frac{5}{7}$ rounds to ______ × ______ = ______

STUDENT PAGE
31
Multiplication Problem Solving

High-Interest Problem Solving

On the board, write a real-life problem that your students can relate to.

EXAMPLE

A car's gas tank holds $7\frac{1}{2}$ gallons. If the tank is $\frac{1}{5}$ full, how many gallons are in the tank?

After students solve the problem, substitute different fractions and mixed numbers. Have students use as many math words as they can. This provides more interesting practice than simply assigning a set of multiplication problems from a textbook or worksheet.

STUDENT PAGE
37
Understanding Division of Fractions

Pictorial Models of Fractions Division

Use pictorial models to show a whole number divided by a fraction. This will help students visualize the process and see why division of fractions results in an answer larger than either of the two numbers.

EXAMPLE $2 \div \frac{1}{4} = 8$

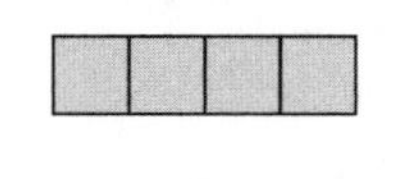
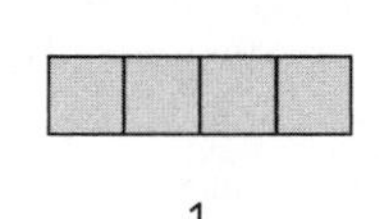
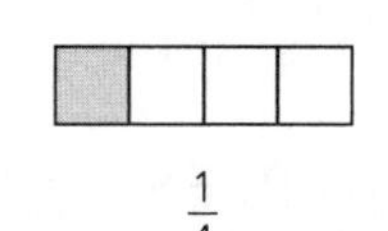

Count: How many fourths are in 2?

1 + 1 ÷ $\frac{1}{4}$

Have students sketch circles to show: $3 \div \frac{1}{4}$, $4 \div \frac{1}{3}$, etc.

Using a ruler will help some students visualize this concept.

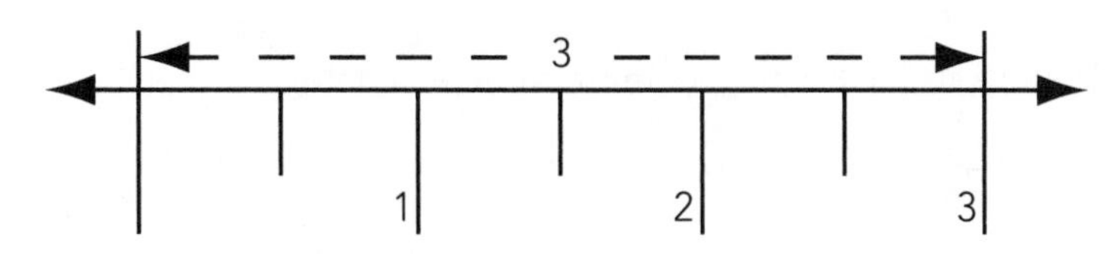

Have students determine how many $\frac{1}{2}$s are in 3.

FRACTIONS: Multiplication and Division

STUDENT PAGE
40
Understanding Division of Fractions

Model Dividing a Fraction by a Fraction

Use graph paper or lined notebook paper turned on its side to show a fraction divided by a fraction.

EXAMPLE $\frac{8}{10} \div \frac{1}{5} = 4$

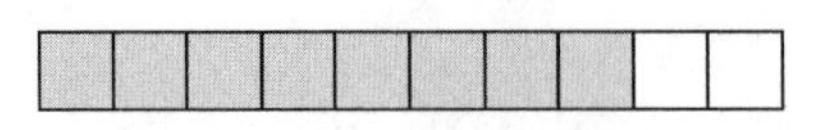

$\frac{8}{10}$ shaded

four $\frac{1}{5}$-regions shaded

Compare the shaded regions. Have students determine how many $\frac{1}{5}$s are in $\frac{8}{10}$.

STUDENT PAGE
52
Division

Using Estimates for Dividing Fractions and Mixed Numbers

Have students practice estimating answers. Tell them to round each mixed number or fraction to the nearest whole number. Remind them that a fraction of $\frac{1}{2}$ or more rounds to the next highest whole number. Then they can divide the whole numbers for a reasonable estimate. Finally, they determine if the estimate is a reasonable answer.

EXAMPLES

A $3\frac{2}{3} \div \frac{5}{6} = ?$

Rounds to $4 \div 1 = 4$ (a reasonable estimate)

B $15\frac{3}{8} \div 3\frac{1}{9} = ?$

Rounds to $15 \div 3 = 5$ (a reasonable estimate)

C $13\frac{5}{6} \div 7\frac{4}{9} = ?$

Rounds to ______ ÷ ______ = ______ (a reasonable estimate)

In real-life, it is often sufficient to round fractions to whole numbers and then divide.

STUDENT PAGE
53
Division

Seeing the Division Factors

Have students work backward from the answer to see the factors in a division problem. You will need to model this a few times so that students will remember to invert the divisor and multiply.

EXAMPLES

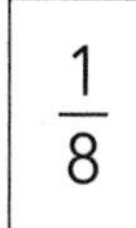

$\frac{1}{8}$	6	$\frac{1}{2}$	$\frac{1}{3}$

Use any 2 of the numbers above to make a true number sentence.

A $\frac{\square}{\square} \div \frac{\square}{\square} = 4$

B $6 \div \frac{\square}{\square} = 12$

C $\frac{\square}{\square} \div \frac{\square}{\square} = 18$

D $\frac{1}{8} \div \frac{\square}{\square} = \frac{1}{4}$

STUDENT PAGE
53
Life-Skills Math

Decrease the Recipe

Cut out recipes from newspapers and magazines and have students figure what amount of each ingredient would be used if the recipe were cut by $\frac{1}{2}$, $\frac{1}{3}$, and $\frac{1}{4}$. Also, give students different situations and have them determine the fraction by which they should decrease a recipe. The same activity can be adapted for fraction multiplication.

Pretest Answer Key

1. $\frac{11}{4}$
2. $\frac{30}{5}$
3. $7.00 or $7
4. $\frac{4}{15}$
5. $6\frac{2}{3}$
6. $2\frac{1}{2}$
7. 42
8. $\frac{8}{5}$
9. 32
10. 3
11. $\frac{8}{9}$
12. $\frac{1}{4}$
13. $\frac{6}{7}$
14. $2\frac{2}{3}$
15. 4 logs
16. $6
17. $3\frac{1}{3}$cups
18. 16 cans
19. $\frac{1}{4}$ pound
20. $10\frac{1}{2}$ yards

Pretest Evaluation Chart

If a student misses any problems in a skill area, assign the practice pages for that skill. However, students may need to complete all practice pages to reinforce areas of weakness.

Skill Area	Pretest Problem #	Skill Section	Review Page
Multiplication	1, 2, 3, 4, 5, 6, 7	7–28	15, 29, 55, 56
Division	8, 9, 10, 11, 12, 13, 14	37–53	54, 55, 56
Multiplication Word Problems	16, 17, 20	30–35 57–68	36 69
Division Word Problems	15, 18, 19	57–68	69
Life-Skills Math	All	70–73	74

Page 7: Fractions of a Set

1. a) $\frac{1}{3}$ $\frac{1}{3}$ $\frac{1}{3}$

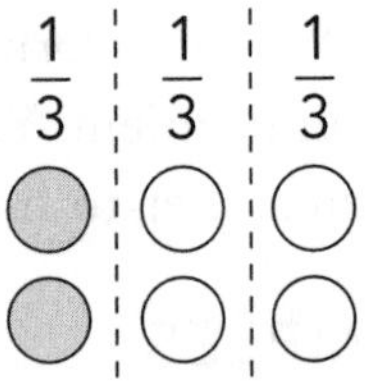

   b) $\frac{1}{3}$ of 6 = 2

2. a) $\frac{1}{5}$ $\frac{1}{5}$ $\frac{1}{5}$ $\frac{1}{5}$ $\frac{1}{5}$

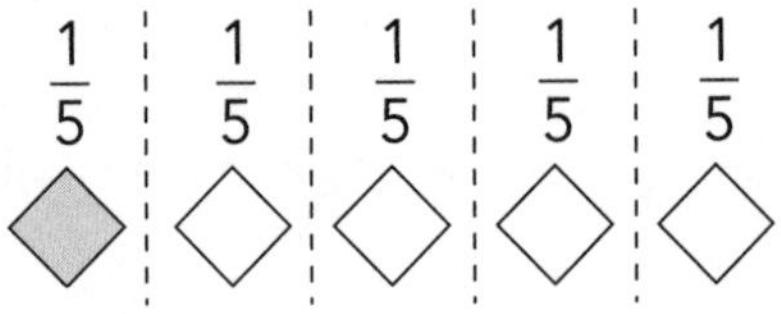

   b) $\frac{1}{5}$ of 5 = 1

3. a) $\frac{1}{2}$ $\frac{1}{2}$

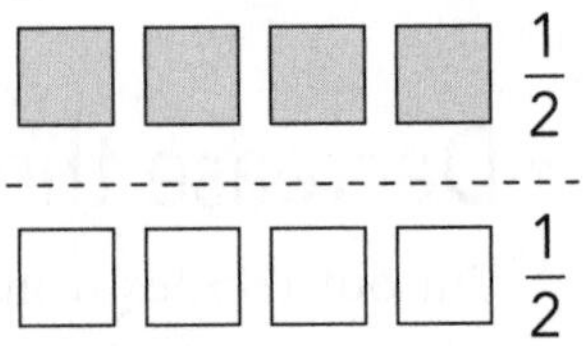

   b) $\frac{1}{2}$ of 8 = 4

4. a)

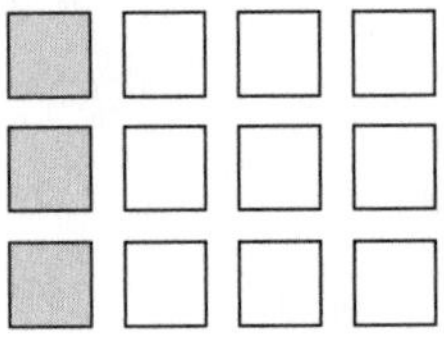

   b) $\frac{1}{4}$ of 12 = 3

5. a)

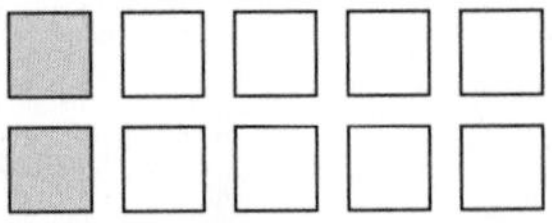

   b) $\frac{1}{5}$ of 10 = 2

6. a)

   b) $\frac{1}{3}$ of 9 = 3

Page 8: Finding One Part of a Set

1.
2. yes
3. $5.00
4. $10.00
5. $2.50
6.
7. yes
8. $3.00
9. $10.00
10. $1.00
11. $7.00
12. $2.00

Page 9: Practice Helps

1. 2	6. 8	11. 4	16. 10
2. 4	7. 9	12. 4	17. 6
3. 1	8. 10	13. 4	18. 2
4. 2	9. 8	14. 3	19. 7
5. 1	10. 5	15. 2	20. 1

Page 10: Shade Fractions of the Sets

1. a) $\frac{1}{3}$ $\frac{1}{3}$ $\frac{1}{3}$

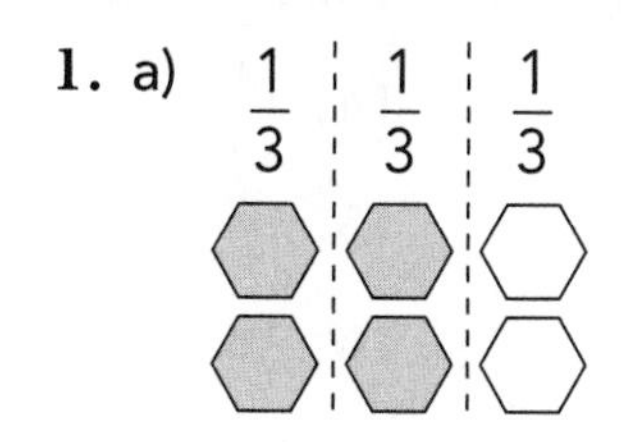

b) $\frac{2}{3}$ of 6 = 4

2. a) $\frac{1}{4}$ $\frac{1}{4}$ $\frac{1}{4}$ $\frac{1}{4}$

b) $\frac{3}{4}$ of 12 = 9

3. a) $\frac{1}{5}$ $\frac{1}{5}$ $\frac{1}{5}$ $\frac{1}{5}$ $\frac{1}{5}$

b) $\frac{4}{5}$ of 10 = 8

4. a)

b) $\frac{3}{4}$ of 8 = 6

5. a)

b) $\frac{3}{5}$ of 10 = 6

6. a)

b) $\frac{3}{4}$ of 4 = 3

Page 11: Finding a Fraction of a Set

1. a) $\frac{1}{4}$ $\frac{1}{4}$ $\frac{1}{4}$ $\frac{1}{4}$

b) If $\frac{1}{4}$ of 8 = 2, then $\frac{3}{4}$ of 8 = 6.

2. a) $\frac{1}{3}$ $\frac{1}{3}$ $\frac{1}{3}$

b) If $\frac{1}{3}$ of 6 = 2, then $\frac{2}{3}$ of 6 = 4.

3. a) $\frac{1}{5}$ $\frac{1}{5}$ $\frac{1}{5}$ $\frac{1}{5}$ $\frac{1}{5}$

b) If $\frac{1}{5}$ of 10 = 2, then $\frac{4}{5}$ of 10 = 8.

Page 12: Apply Your Skills

1. If $\frac{1}{4}$ of 24 = 6, then $\frac{3}{4}$ of 24 = 18.
2. If $\frac{1}{5}$ of 15 = 3, then $\frac{3}{5}$ of 15 = 9.
3. If $\frac{1}{5}$ of 20 = 4, then $\frac{4}{5}$ of 20 = 16.
4. If $\frac{1}{9}$ of 45 = 5, then $\frac{5}{9}$ of 45 = 25.
5. If $\frac{1}{7}$ of 63 = 9, then $\frac{6}{7}$ of 63 = 54.
6. If $\frac{1}{8}$ of 48 = 6, then $\frac{7}{8}$ of 48 = 42.
7. If $\frac{1}{5}$ of 5 = 1, then $\frac{4}{5}$ of 5 = 4.
8. If $\frac{1}{10}$ of 100 = 10, then $\frac{7}{10}$ of 100 = 70.

Page 13: Multiplication Is Repeated Addition

1. $\frac{1}{2} + \frac{1}{2} + \frac{1}{2} + \frac{1}{2} + \frac{1}{2} = \frac{5}{2} = 2\frac{1}{2}$
2. $\frac{2}{5} + \frac{2}{5} + \frac{2}{5} = \frac{6}{5} = 1\frac{1}{5}$
3. $\frac{3}{4} + \frac{3}{4} + \frac{3}{4} + \frac{3}{4} = \frac{12}{4} = 3$
4. $\frac{1}{2} + \frac{1}{2} + \frac{1}{2} = \frac{3}{2} = 1\frac{1}{2}$

Page 14: Multiplication Models

1. 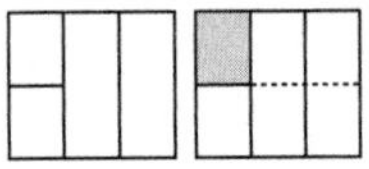

$\frac{1}{2}$ of $\frac{1}{3} = \frac{1}{6}$

$\frac{1}{2} \times \frac{1}{3} = \frac{1}{6}$

2. 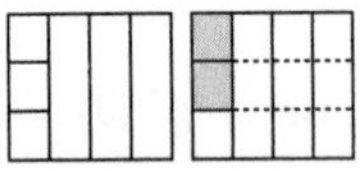

$\frac{2}{3}$ of $\frac{1}{4} = \frac{2}{12}$

$\frac{2}{3} \times \frac{1}{4} = \frac{2}{12}$

3. 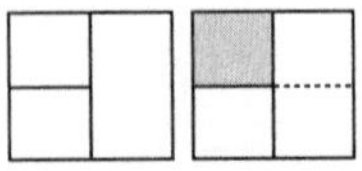

$\frac{1}{2}$ of $\frac{1}{2} = \frac{1}{4}$

$\frac{1}{2} \times \frac{1}{2} = \frac{1}{4}$

4. 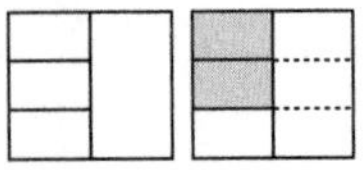

$\frac{2}{3}$ of $\frac{1}{2} = \frac{2}{6}$

$\frac{2}{3} \times \frac{1}{2} = \frac{2}{6}$

Page 15: Understanding Multiplication of Fractions Review

1. a) 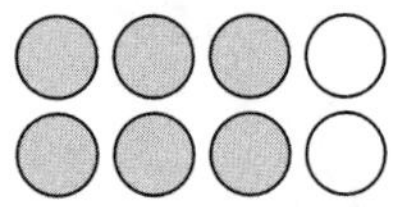

   b) 6
2. a) $18

   b) $16

   c) $7
3. a) 6

   b) 8

   c) 3
4. a) 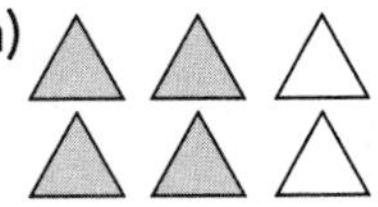

   b) 4
5. a) $2\frac{1}{4}$

   b) $1\frac{2}{3}$
6. a) 25

   b) 30

Page 16: Multiplying Fractions and Whole Numbers

1. $1\frac{1}{4}$
2. $1\frac{3}{5}$
3. $2\frac{1}{2}$
4. $1\frac{3}{5}$
5. $5\frac{3}{5}$
6. 3
7. $5\frac{1}{3}$
8. 2
9. $5\frac{2}{5}$
10. $2\frac{2}{5}$
11. $1\frac{3}{5}$
12. $\frac{2}{3}$
13. $5\frac{1}{4}$
14. $1\frac{1}{8}$
15. $1\frac{7}{8}$
16. 18

Page 17: A Fraction Times a Fraction

1. $\frac{1}{24}$
2. $\frac{5}{18}$
3. $\frac{1}{8}$
4. $\frac{1}{6}$
5. $\frac{1}{5}$
6. $\frac{3}{8}$
7. $\frac{1}{2}$
8. $\frac{2}{5}$
9. $\frac{2}{5}$
10. $\frac{1}{20}$
11. $\frac{5}{8}$
12. $\frac{1}{4}$
13. $\frac{2}{21}$
14. $\frac{1}{3}$

Page 18: Simplify the Fractions

1. $\frac{5}{\underset{3}{\cancel{6}}} \times \frac{\overset{2}{\cancel{4}}}{7}$
2. $\frac{\overset{1}{\cancel{5}}}{9} \times \frac{2}{\underset{3}{\cancel{15}}}$
3. $\frac{1}{\underset{1}{\cancel{2}}} \times \frac{\overset{3}{\cancel{6}}}{7}$
4. $\frac{1}{\underset{4}{\cancel{12}}} \times \frac{\overset{1}{\cancel{3}}}{5}$
5. $\frac{\overset{3}{\cancel{12}}}{1} \times \frac{3}{\underset{1}{\cancel{4}}}$
6. $\frac{3}{\underset{2}{\cancel{8}}} \times \frac{\overset{3}{\cancel{12}}}{5}$
7. $\frac{1}{4} \times \frac{\overset{3}{\cancel{9}}}{\underset{5}{\cancel{15}}}$
8. $\frac{\overset{1}{\cancel{4}}}{3} \times \frac{17}{\underset{2}{\cancel{8}}}$
9. $\frac{\overset{1}{\cancel{2}}}{3} \times \frac{7}{\underset{4}{\cancel{8}}}$
10. $\frac{9}{\underset{1}{\cancel{10}}} \times \frac{\overset{1}{\cancel{10}}}{11}$
11. $\frac{4}{\underset{1}{\cancel{5}}} \times \frac{\overset{3}{\cancel{15}}}{17}$
12. $\frac{\overset{2}{\cancel{8}}}{9} \times \frac{1}{\underset{1}{\cancel{4}}}$

Page 19: Simplify First

1. $\frac{\overset{1}{\cancel{4}}}{\underset{3}{\cancel{9}}} \times \frac{\overset{1}{\cancel{3}}}{\underset{2}{\cancel{8}}}$
2. $\frac{\overset{3}{\cancel{18}}}{\underset{1}{\cancel{5}}} \times \frac{\overset{1}{\cancel{5}}}{\underset{1}{\cancel{6}}}$
3. $\frac{\overset{1}{\cancel{5}}}{\underset{6}{\cancel{18}}} \times \frac{\overset{1}{\cancel{3}}}{\underset{4}{\cancel{20}}}$
4. $\frac{\overset{3}{\cancel{6}}}{\underset{5}{\cancel{25}}} \times \frac{\overset{1}{\cancel{5}}}{\underset{4}{\cancel{8}}}$
5. $\frac{\overset{1}{\cancel{2}}}{\underset{1}{\cancel{3}}} \times \frac{\overset{5}{\cancel{15}}}{\underset{8}{\cancel{16}}}$
6. $\frac{\overset{1}{\cancel{4}}}{\underset{1}{\cancel{5}}} \times \frac{\overset{5}{\cancel{25}}}{\underset{8}{\cancel{32}}}$
7. $\frac{\overset{2}{\cancel{14}}}{\underset{3}{\cancel{15}}} \times \frac{\overset{4}{\cancel{20}}}{\underset{3}{\cancel{21}}}$
8. $\frac{\overset{2}{\cancel{8}}}{\underset{3}{\cancel{9}}} \times \frac{\overset{1}{\cancel{3}}}{\underset{1}{\cancel{4}}}$
9. $\frac{\overset{1}{\cancel{32}}}{\underset{5}{\cancel{35}}} \times \frac{\overset{3}{\cancel{21}}}{\underset{1}{\cancel{32}}}$
10. $\frac{\overset{5}{\cancel{35}}}{\underset{3}{\cancel{48}}} \times \frac{\overset{1}{\cancel{16}}}{\underset{3}{\cancel{21}}}$
11. $\frac{\overset{1}{\cancel{11}}}{\underset{3}{\cancel{12}}} \times \frac{\overset{1}{\cancel{4}}}{\underset{3}{\cancel{33}}}$
12. $\frac{\overset{3}{\cancel{24}}}{\underset{5}{\cancel{25}}} \times \frac{\overset{3}{\cancel{15}}}{\underset{4}{\cancel{32}}}$

Page 20: Simplify Before Multiplying

1. $\frac{1}{8}$
2. $\frac{1}{4}$
3. $\frac{4}{7}$
4. $\frac{2}{7}$
5. $12\frac{1}{4}$
6. $\frac{1}{3}$
7. $\frac{2}{3}$
8. $4\frac{4}{5}$
9. $\frac{7}{10}$
10. $\frac{3}{14}$
11. 2
12. $\frac{1}{6}$
13. $\frac{1}{6}$
14. $\frac{1}{10}$
15. $\frac{1}{3}$

Page 21: Practice Your Skills

1. $\frac{5}{14}$
2. $3\frac{1}{3}$
3. $\frac{1}{9}$
4. $1\frac{2}{5}$
5. $\frac{3}{10}$
6. $\frac{1}{3}$
7. $\frac{2}{5}$
8. $\frac{1}{2}$
9. $\frac{1}{9}$
10. 6
11. $2\frac{1}{2}$
12. $\frac{1}{2}$
13. $3\frac{1}{3}$
14. $4\frac{1}{5}$
15. $\frac{3}{25}$
16. $\frac{5}{8}$
17. $3\frac{3}{4}$
18. $\frac{1}{6}$
19. $\frac{1}{3}$
20. $\frac{2}{15}$

Page 22: Fractions Equal to Whole Numbers

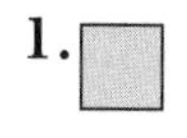
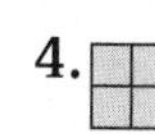
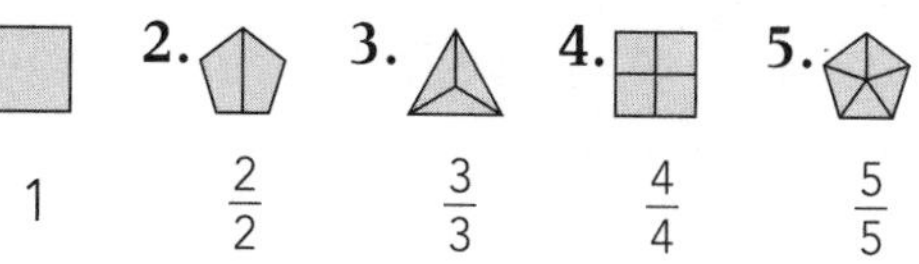

1. 1
2. $\frac{2}{2}$
3. $\frac{3}{3}$
4. $\frac{4}{4}$
5. $\frac{5}{5}$
6. $1 = \frac{2}{2} = \frac{3}{3} = \frac{4}{4} = \frac{5}{5} = \frac{6}{6} = \frac{7}{7} = \frac{8}{8} = \frac{9}{9}$

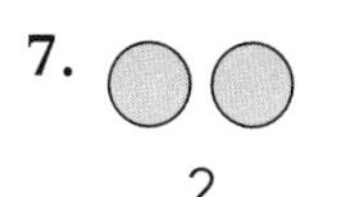
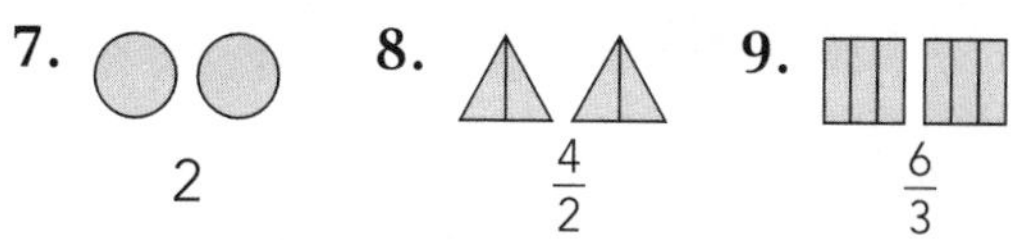

7. 2
8. $\frac{4}{2}$
9. $\frac{6}{3}$
10. $\frac{8}{4}$
11. $\frac{10}{5}$
12. $2 = \frac{4}{2} = \frac{6}{3} = \frac{8}{4} = \frac{10}{5} = \frac{12}{6} = \frac{14}{7} = \frac{16}{8} = \frac{18}{9}$

FRACTIONS: Multiplication and Division

Page 23: Changing Whole Numbers to Fractions

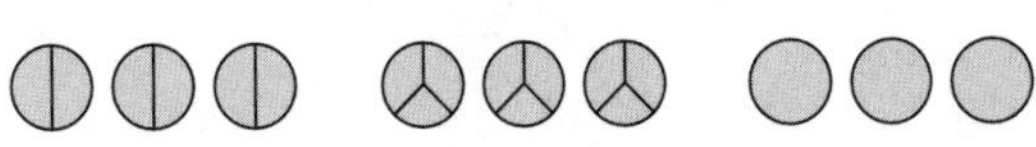
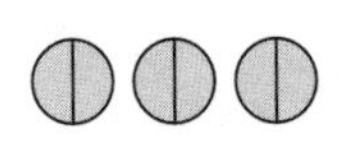
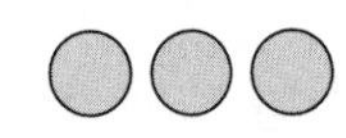

1. $\frac{6}{2}$ **2.** $\frac{9}{3}$ **3.** 3

4. $3 = \frac{6}{2} = \frac{9}{3} = \frac{12}{4} = \frac{15}{5} = \frac{18}{6} = \frac{21}{7} = \frac{24}{8}$

5. $4 = \frac{8}{12} = \frac{12}{3} = \frac{16}{4} = \frac{20}{5} = \frac{24}{6} = \frac{28}{7} = \frac{32}{8}$

6. $1 = \frac{8}{8}$ (1×8)

7. $3 = \frac{9}{3}$ (3×3)

8. $4 = \frac{40}{10}$ (4×10)

9. $8 = \frac{32}{4}$ (8×4)

10. $5 = \frac{30}{6}$ (5×6)

11. $9 = \frac{27}{3}$ (9×3)

12. $4 = \frac{32}{8}$ (4×8)

13. $6 = \frac{60}{10}$ (6×10)

14. $7 = \frac{49}{7}$ (7×7)

15. $7 = \frac{21}{3}$ (7×3)

16. $15 = \frac{30}{2}$ (15×2)

17. $7 = \frac{35}{5}$ (7×5)

18. $9 = \frac{36}{4}$ (9×4)

19. $8 = \frac{24}{3}$ (8×3)

20. $4 = \frac{16}{4}$ (4×4)

Page 24: Change Mixed Numbers to Improper Fractions

1. $3\frac{3}{4} = \frac{15}{4}$

2. $1\frac{2}{3} = \frac{5}{3}$

3. $2\frac{5}{6} = \frac{17}{6}$

4. $5\frac{2}{3} = \frac{17}{3}$ Think: $5\frac{2}{3} = \overset{(5\times3)}{\frac{15}{3}} + \frac{2}{3} = \frac{17}{3}$

5. $6\frac{2}{5} = \frac{32}{5}$ Think: $6\frac{2}{5} = \overset{(6\times5)}{\frac{30}{5}} + \frac{2}{5} = \frac{32}{5}$

6. $5\frac{1}{2} = \frac{11}{2}$ Think: $5\frac{1}{2} = \overset{(5\times2)}{\frac{10}{2}} + \frac{1}{2} = \frac{11}{2}$

7. $3\frac{5}{8} = \frac{29}{8}$ Think: $3\frac{5}{8} = \overset{(3\times8)}{\frac{24}{8}} + \frac{5}{8} = \frac{29}{8}$

Page 25: More Practice Changing to Improper Fractions

1. $\frac{5}{2}$
2. $\frac{41}{6}$
3. $\frac{14}{9}$
4. $\frac{9}{4}$
5. $\frac{51}{7}$
6. $\frac{29}{3}$
7. $\frac{28}{3}$
8. $\frac{21}{8}$
9. $\frac{23}{4}$
10. $\frac{47}{6}$
11. $\frac{26}{5}$
12. $\frac{23}{6}$
13. $\frac{11}{3}$
14. $\frac{25}{2}$
15. $\frac{39}{4}$
16. $\frac{47}{3}$
17. $\frac{33}{8}$
18. $\frac{42}{5}$

Page 26: Rename the Mixed Number

1. $\frac{11}{21}$
2. $2\frac{4}{5}$
3. $\frac{33}{50}$
4. $3\frac{1}{5}$
5. 2
6. $\frac{1}{2}$
7. $1\frac{1}{2}$
8. $2\frac{1}{6}$
9. $2\frac{1}{8}$
10. $1\frac{1}{35}$

Page 27: Multiplying Mixed Numbers

1. $12\frac{2}{3}$
2. $3\frac{5}{8}$
3. 8
4. $9\frac{1}{5}$
5. 24
6. $7\frac{2}{5}$
7. $7\frac{2}{9}$
8. $6\frac{4}{7}$

Page 28: Master the Skills

1. $1\frac{2}{5}$
2. $\frac{3}{5}$
3. $2\frac{1}{2}$
4. 10
5. $\frac{2}{15}$
6. $\frac{1}{4}$
7. $\frac{3}{5}$
8. $1\frac{11}{12}$
9. $1\frac{1}{14}$
10. $\frac{11}{21}$
11. $6\frac{1}{4}$
12. 14
13. $\frac{2}{3}$
14. $1\frac{5}{8}$
15. $6\frac{2}{3}$
16. $5\frac{1}{4}$

Page 29: Multiplication Review

1. \$5.00
2. 5
3. a) $\frac{1}{4}$ $\frac{1}{4}$ $\frac{1}{4}$ $\frac{1}{4}$

   b) If $\frac{1}{4}$ of 8 = 2, then $\frac{3}{4}$ of 8 = 6.
4. If $\frac{1}{4}$ of 36 = 9, then $\frac{3}{4}$ of 36 = 27.
5. $1\frac{1}{5}$
6. $\times$
7. $2\frac{2}{5}$
8. $\frac{7}{24}$
9. $\frac{3}{17}$
10. $\frac{2}{3}$
11. $\frac{1}{9}$
12. 4
13. $\frac{12}{6}$
14. $\frac{28}{4}$
15. $\frac{23}{5}$
16. $\frac{49}{8}$
17. 2
18. 28

Page 30: Find a Fraction of an Amount

1. $21\frac{1}{3}$
2. 3
3. \$33.00
4. 3

Page 31: Does the Answer Make Sense?

1. $\frac{1}{2} \times 5 = 2\frac{1}{2}$; $2\frac{1}{2}$
2. $2\frac{2}{3} \times \$3 = \$8$; 8
3. $24 \times \frac{3}{4} = 18$; 18
4. $6\frac{1}{2} \times 10 = 65$; 65
5. $\frac{1}{5} \times \$145 = \$29$; 29
6. $3\frac{3}{4} \times 6 = 22\frac{1}{2}$; $22\frac{1}{2}$

Page 32: Number Sentences

1. $\frac{1}{2} \times 4 = 2$; 2
2. $\frac{3}{4} \times 8 = 6$; 6
3. $\frac{3}{4} \times 4 = 3$; 3
4. $16\frac{1}{2} \times \frac{1}{3} = 5\frac{1}{2}$; $5\frac{1}{2}$
5. $\frac{5}{8} \times 48 = 30$; 30
6. $\frac{3}{4} \times 28 = 21$; 21
7. $45 \times \frac{1}{3} = \$15.00$; 15.00
8. $6\frac{1}{2} \times 6 = 39$; 39

Page 33: Think It Through

Answers should be similar to these.

1. a) How much sugar is used in all?
   b) How much sugar is used if Margo doubles the recipe?
2. a) How much do 2 boxes weigh?
   b) How much do 3 boxes weigh?
3. a) How much did he save?
   b) What was the sale price of the jacket?
4. a) How many hours are in 2 shifts?
   b) How many hours are in 3 shifts?
5. a) How much did he save?
   b) What was the sale price of the backpack?
6. a) How many slices did she eat?
   b) How many slices were left over?

Page 34: Write a Question

1. a) How much money was saved?
   b) $34 \times \frac{1}{7} = \$5$
2. a) How many pounds were eaten?
   b) $\frac{1}{2} \times 15 = 7\frac{1}{2}$
3. a) How long is the string in centimeters?
   b) $\frac{2}{5} \times 17 = 6\frac{4}{5}$
4. a) How long are the 5 Dachshunds?
   b) $1\frac{1}{3} \times 5 = 6\frac{2}{3}$
5. a) How much does she save each week?
   b) $\frac{2}{5} \times 456 = \$304$
6. a) How far does she walk in one week?
   b) $2\frac{1}{4} \times 6 = 13\frac{1}{2}$

Page 35: Practice Helps

1. $337\frac{1}{2}$ miles
2. $10.67
3. $63\frac{3}{4}$ pounds
4. 36 hours
5. 8 pounds
6. $1\frac{9}{16}$ inches
7. 28 cups
8. $72

Page 36: Multiplication Problem-Solving Review

1. 352 ounces
2. $221\frac{1}{3}$ miles
3. $17\frac{1}{2}$
4. $1\frac{3}{4}$
5. $11
6. $38.25
7. $132
8. $2\frac{1}{3}$ pounds

Page 37: Divide Whole Numbers by Fractions

1. a) 10
   b) 10
2. a) 12
   b) 12
3. a) 9
   b) 9
4. a) 8
   b) 8
5. a) 20
   b) 20
6. a) 24
   b) 24

Page 38: Think About Fraction Division

1. $2 \div \frac{1}{4} = 8$
2. $2 \times 4 = 8$
3. $3 \div \frac{1}{2} = 6$
4. $3 \times 2 = 6$
5. 4
6. 4
7. 6
8. 12
9. 10
10. 20
11. 8
12. 9
13. 15
14. 12
15. 16
16. 8

Page 39: Reciprocals

	Reciprocal
1.	$\frac{5}{3}$
2.	$\frac{8}{3}$
3.	$\frac{2}{8}$
4.	$\frac{4}{3}$
5.	$\frac{3}{5}$

	Number		Reciprocal
11.	$3\frac{1}{2}$	$= \frac{7}{2}$	$\frac{2}{7}$
12.	$2\frac{3}{4}$	$= \frac{11}{4}$	$\frac{4}{11}$
13.	$4\frac{1}{5}$	$= \frac{21}{5}$	$\frac{5}{21}$
14.	$3\frac{5}{6}$	$= \frac{23}{6}$	$\frac{6}{23}$
15.	$1\frac{3}{5}$	$= \frac{8}{5}$	$\frac{5}{8}$

	Number	Reciprocal
6.	$5 = \frac{5}{1}$	$\frac{1}{5}$
7.	$7 = \frac{7}{1}$	$\frac{1}{7}$
8.	$3 = \frac{3}{1}$	$\frac{1}{3}$
9.	$4 = \frac{4}{1}$	$\frac{1}{4}$
10.	$15 = \frac{15}{1}$	$\frac{1}{15}$

Page 40: Dividing with Fractions

1. a) 6
   b) 6
   c) $\frac{8}{1}$
2. a) 4
   b) 4
   c) $\frac{6}{1}$
3. a) 2
   b) 2
   c) $\frac{4}{1}$
4. a) 4
   b) 4
   c) $\frac{8}{1}$

Page 41: Understanding Division of Fractions Review

1. a) 16
   b) 16
2. a) 24
   b) 20
3. a) $\frac{4}{3}$
   b) $\frac{8}{5}$
   c) $\frac{3}{2}$
4. a) $\frac{5}{11}$
   b) $\frac{3}{14}$
   c) $\frac{8}{55}$
5. $2\frac{1}{2}$
6. $\frac{2}{3}$
7. $\frac{2}{3}$
8. $1\frac{3}{7}$

Page 42: Multiply by the Reciprocal

A. $\frac{8}{9}$

B. $\frac{16}{35}$

1. $\frac{4}{5}$
2. $\frac{3}{8}$
3. $\frac{2}{3} \times \frac{8}{7} = \frac{16}{21}$
4. $\frac{1}{8} \times \frac{3}{2} = \frac{3}{16}$
5. $\frac{4}{5} \times \frac{8}{7} = \frac{32}{35}$
6. $\frac{1}{5} \times \frac{3}{1} = \frac{3}{5}$
7. $\frac{8}{35}$
8. $\frac{24}{35}$
9. $\frac{1}{2} \times \frac{5}{3} = \frac{5}{6}$
10. $\frac{2}{5} \times \frac{4}{3} = \frac{8}{15}$
11. $\frac{3}{4} \times \frac{5}{4} = \frac{15}{16}$
12. $\frac{1}{6} \times \frac{5}{1} = \frac{5}{6}$

Page 43: A Fraction Divided by a Fraction

1. $\frac{1}{5}$
2. $1\frac{1}{4}$
3. $\frac{3}{8}$
4. $1\frac{1}{3}$
5. $\frac{4}{5}$
6. $2\frac{2}{5}$
7. $\frac{3}{4}$
8. $1\frac{7}{8}$
9. $\frac{1}{2}$
10. $3\frac{1}{3}$
11. $1\frac{1}{15}$
12. 3

Page 44: Dividing by a Fraction

1. a) 6
b) 6

2. a) 5
b) 5

3. a) 3
b) 3

4. a) 3
b) 3

5. a) 2
b) 2

Page 45: Think About Dividing by Fractions

1. a) 3
b) 3

2. a) 5
b) 5

3. a) 7
b) 7

4. a) 9
b) 9

5. a) 2
b) 2

Page 46: Using Drawings

A. 5

B. 5

1. a) 3
b) 3

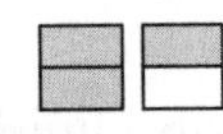

2. a) 7
b) 7

3. a) 10
b) 10

4. a) 14
b) 14

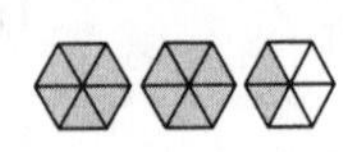

Page 47: Dividing a Mixed Number by a Fraction

1. $3\frac{3}{4}$

2. $3\frac{5}{9}$

3. $7\frac{1}{2}$

4. $5\frac{1}{4}$

5. $2\frac{2}{5}$

6. 30

7. $13\frac{1}{3}$

8. $5\frac{3}{5}$

9. 5

10. 2

11. 25

12. $12\frac{3}{4}$

Page 48: Mixed Practice

1. 16

2. $\frac{6}{7}$

3. 14

4. $\frac{1}{3}$

5. $17\frac{1}{2}$

6. 6

7. 25

8. 7

9. 10

10. $1\frac{1}{2}$

11. 4

12. 10

13. 4

14. $9\frac{1}{3}$

15. 3

16. $1\frac{1}{2}$

Page 49: Divide Whole Numbers by Fractions

1. 27

2. 12

3. $17\frac{1}{2}$

4. 18

5. 21

6. 20

7. 6

8. 22

9. 12

10. 49

11. 15

12. $7\frac{1}{5}$

Page 50: Divide by Whole Numbers

1. $\frac{7}{12}$
2. $\frac{3}{8}$
3. $\frac{3}{25}$
4. $\frac{2}{5}$
5. $1\frac{2}{3}$
6. $2\frac{1}{5}$
7. $\frac{6}{7}$
8. $\frac{7}{8}$
9. $2\frac{1}{2}$
10. $\frac{5}{9}$
11. $1\frac{1}{3}$
12. $1\frac{3}{4}$

Page 51: Divide by Mixed Numbers

1. $\frac{1}{4}$
2. $\frac{1}{5}$
3. $\frac{1}{16}$
4. $\frac{6}{25}$
5. $\frac{3}{20}$
6. $\frac{7}{30}$
7. $\frac{2}{3}$
8. $\frac{1}{6}$
9. $\frac{18}{49}$
10. $\frac{3}{37}$
11. $\frac{8}{27}$
12. $\frac{2}{11}$

Page 52: Divide Two Mixed Numbers

1. $1\frac{7}{8}$
2. 2
3. $\frac{3}{5}$
4. 3
5. $\frac{13}{16}$
6. $\frac{13}{19}$
7. $1\frac{2}{7}$
8. $2\frac{8}{11}$
9. 7
10. $3\frac{3}{4}$
11. $2\frac{2}{5}$
12. $5\frac{5}{6}$

Page 53: Division Practice

1. 6
2. $\frac{15}{22}$
3. $\frac{8}{27}$
4. $3\frac{3}{4}$
5. $2\frac{7}{9}$
6. 6
7. $\frac{1}{10}$
8. $1\frac{1}{3}$
9. $1\frac{7}{15}$
10. $2\frac{1}{2}$

Page 54: Division Review

1. 8
2. 15
3. $\frac{5}{19}$
4. $1\frac{1}{2}$
5. $2\frac{2}{3}$
6. 5
7. $4\frac{2}{7}$
8. 6
9. 55
10. $\frac{5}{8}$
11. $\frac{3}{5}$
12. 2

Page 55: Use All Operations

1. $2\frac{2}{5}$
2. $\frac{11}{16}$
3. $3\frac{4}{5}$
4. $\frac{5}{8}$
5. $7\frac{2}{3}$
6. 9
7. $14\frac{5}{12}$
8. $4\frac{3}{4}$
9. $1\frac{1}{15}$
10. 15
11. $8\frac{1}{4}$
12. $\frac{3}{14}$

Page 56: Putting It All Together

1. $\frac{1}{4} < 3\frac{1}{3}$
2. $3\frac{1}{3} > 3\frac{1}{11}$
3. $5\frac{5}{18} > 4\frac{7}{10}$
4. $4\frac{1}{2} > 3\frac{1}{23}$
5. $5\frac{7}{8} > 5\frac{3}{8}$
6. $\frac{1}{8} < \frac{3}{8}$
7. $5\frac{7}{8} > 4\frac{3}{8}$
8. $\frac{13}{16} < 1\frac{1}{2}$

Page 57: Does the Answer Make Sense?

1. $4 \div \frac{1}{4} = 16$; 16
2. $1\frac{1}{4} \times 3\frac{3}{4} = 4\frac{11}{16}$; $4\frac{11}{16}$
3. $12 \div \frac{3}{4} = 16$; 16
4. $1\frac{1}{4} \times 7\frac{1}{2} = 9\frac{3}{8}$; $9\frac{3}{8}$
5. $15 \div \frac{3}{4} = 20$; 20
6. $6 \div \frac{2}{5} = 15$; 15

Page 58: Decide to Multiply or Divide

1. $\frac{1}{2} \times 5 = 2\frac{1}{2}$; $2\frac{1}{2}$
2. $35 \times \frac{1}{5} = \$7.00$; 7.00
3. $5\frac{1}{2} \div 2 = 2\frac{3}{4}$; $2\frac{3}{4}$
4. $5 \div \frac{1}{2} = 10$; 10

Page 59: Mixed Multiplication and Division

1. $3\frac{1}{2} \div \frac{1}{4} = 14$; 14
2. $\frac{1}{5} \times 7\frac{1}{2} = 1\frac{1}{2}$; $1\frac{1}{2}$
3. $3\frac{1}{3} \times 2\frac{1}{4} = 7\frac{1}{2}$; $7\frac{1}{2}$
4. $1\frac{1}{2} \div 2 = \frac{3}{4}$; $\frac{3}{4}$
5. $10 \div 2\frac{1}{2} = 4$; 4
6. $5 \times 8\frac{1}{2} = 42\frac{1}{2}$; $42\frac{1}{2}$
7. $25 \div 2\frac{1}{2} = 10$; 10
8. $\frac{3}{4} \times 2 = 1\frac{1}{2}$; $1\frac{1}{2}$

Page 60: Think It Through

Answers should be similar to these.

1. a) How many cups of flour and milk are used in all?
   b) How many more cups of flour were used?
2. a) How much was marked off?
   b) What was the sale price of the sweater?
3. a) How many hours did Bob work in two $8\frac{1}{2}$-hour shifts?
   b) If Bob worked three $8\frac{1}{2}$-hour shifts, how many hours did he work?
4. a) About how many miles did Bonita travel each hour?
   b) If she travels at the same speed, about how many miles would Bonita travel in 5 hours?
5. a) 5 centimeters are about how many inches?
   b) 3 centimeters are about how many inches?
6. a) How much does each box weigh?
   b) How much do 5 boxes weigh?

Page 61: Write a Question

Sample questions:

1. How much money was marked off?
   $36 \times \frac{1}{4} = \$9.00$
2. How many miles did he walk in all?
   $14 \times 1\frac{1}{2} = 21$ miles
3. How many $1\frac{1}{2}$-foot pieces will he have?
   $6 \div 1\frac{1}{2} = 4$ pieces
4. What is the total weight of 12 sacks?
   $\frac{1}{2} \times 12 = 6$ pounds
5. About how many centimeters are in 20 inches?
   $20 \div \frac{2}{5} = 50$ centimeters
6. How much does Lynn save each week?
   $125 \times \frac{1}{5} = \$25.00$

Page 62: Apply the Operations

1. D
2. B
3. A
4. C
5. $\frac{1}{2} \times 1\frac{1}{4} = \frac{5}{8}; \frac{5}{8}$
6. $15 \div 3\frac{3}{4} = 4; 4$
7. $1\frac{1}{2} + 3\frac{3}{4} = 5\frac{1}{4}; 5\frac{1}{4}$
8. $8\frac{1}{2} - 1\frac{1}{8} = 7\frac{3}{8}; 7\frac{3}{8}$

Page 63: Choose the Question

1. addition
2. subtraction
3. multiplication
4. multiplication
5. division
6. addition
7. division
8. multiplication
9. subtraction
10. addition

Page 64: Mixed Problem Solving

1. a) $\frac{5}{8}$ of the strawberries were picked
   b) $\frac{3}{8}$ left to pick
2. $3\frac{1}{2} - 1\frac{1}{4} = 2\frac{1}{4}; 2\frac{1}{4}$
3. $3\frac{1}{4} + 2\frac{1}{3} = 5\frac{7}{12}; 5\frac{7}{12}$
4. $\frac{2}{3} \times \frac{1}{2} = \frac{1}{3}; \frac{1}{3}$
5. $\frac{1}{10} + \frac{1}{5} = \frac{3}{10}; \frac{3}{10}$
6. $4 \div \frac{1}{2} = 8; 8$

Page 65: Throw Away Extra Information

1. a) Facts not needed: $4.98; 7 people
   b) You don't need the number of people or the price to figure out number of pounds.
   c) $1\frac{1}{4}$
2. a) Facts not needed: 225 miles; 485 miles
   b) You don't need the number of miles to figure out number of hours.
   c) $12\frac{5}{6}$
3. a) Facts not needed: $36.19; 4 people; 12 friends
   b) You don't need the price and number of people to figure out number of pounds.
   c) $3\frac{3}{4}$

Page 66: Two-Step Story Problems

A. $54 \times \frac{1}{3} = \$18$

B. $\$54 - \$18 = \$36$

1. a) $10 \times 2\frac{1}{2} = 4$ boards
   b) $20 \div 4 = 5$ boards 10 feet long
2. a) $180 \div 60 = 3$ batches
   b) $1\frac{1}{4} \times 3 = 3\frac{3}{4}$ cups
3. a) $40 \times \frac{4}{5} = 32$ questions
   b) $40 \times \frac{7}{8} = 35$ questions
   c) yes
4. a) $60 \times \frac{1}{5} = 12$ cookies
   b) $60 - 12 = 48$ cookies

Page 67: More Two-Step Story Problems

A. $6\frac{3}{4}$ gallons

B. $2\frac{1}{4}$ gallons

1. 4

2. $55

3. 20

4. 4

Page 68: Multistep Word Problems

Q1: $2\frac{1}{4}$ hours

Q2: $2\frac{3}{4}$ hours

Q3: $\frac{1}{2}$ hour

Answers should be similar to these.

1. Q1: How much does he spend on rent? $145

Q2: How much does he spend on food? $87

Q3: After subtracting food and rent, how much does he have left? $203

2. Q1: How long is the first shift? 6 hours

Q2: How long is the second shift? 8 hours

Q3: How many hours of the day are left for the third shift? 10 hours

3. Q1: How long did Walter take altogether? $17\frac{1}{4}$ hours

Q2: How long did Carin take altogether? 18 hours

Q3: How much longer did Carin take? $\frac{3}{4}$ hour

4. Q1: How much flour did she use? 4 lb

Q2: How much flour is left over? $2\frac{1}{2}$ lb

Q3: How many times can she fill the bread recipe with the remaining flour? 2 times

Page 69: Mixed Problem-Solving Review

1. 24 sections

2. $38.33

3. $34\frac{1}{8}$ inches

4. $26\frac{5}{6}$ hours

5. $8

6. a) 16 cookies
b) 64 cookies

7. $1\frac{1}{2}$ cups

8. 15 people

Page 70: Decrease a Recipe

1. a) $\frac{3}{4}$

b) $1\frac{1}{2}$

c) $\frac{1}{8}$

d) $\frac{5}{8}$

e) $\frac{3}{8}$

f) $\frac{1}{6}$

g) 1

2. 6, or $\frac{1}{2}$ dozen

3. a) 12

b) 6 batches

Page 71: Increase a Recipe

1. a) 5
   b) $2\frac{2}{3}$
   c) 2
   d) 1
   e) $1\frac{1}{2}$
   f) 1
   g) 16
   h) 4
2. 2 batches
3. a) 4
   b) $1\frac{1}{2}$
   c) $2\frac{1}{4}$
   d) 6

Page 72: At the Store

Item	Cost per Pound	Cost
$3\frac{1}{2}$ pounds of green beans	1. a) $1.52	b) $5.32
$2\frac{3}{4}$ pounds of pears	2. a) $1.79	b) $4.73
$1\frac{1}{3}$ pounds of carrots	3. a) $.83	b) $1.12
7 pounds of apples	4. a) $1.25	b) $8.75
$1\frac{1}{8}$ pounds of cucumbers	5. a) $2.44	b) $2.70
	Total Cost	6. $22.62

7. $10.24
8. $2.62
9. $2.42
10. $1.32

Page 73: Common Discounts

1. $142.44
2. $26.63
3. $7.49
4. a) $8.00
   b) $4.13
   c) $9.00
   d) $1.33
   e) $49.00
   f) $299.50

Page 74: Life-Skills Math Review

1. $27.96
2. $6.30
3. $3\frac{3}{4}$ tsp
4. $74.90
5. $10.84
6. $1\frac{2}{3}$ cups

Page 75: Cumulative Review

1. a) ●●●●●●○○ / ●●●●●●○○
   b) 12
2. $\frac{1}{3}$
3. $4\frac{1}{5}$
4. $9\frac{1}{15}$
5. Q1. Questions and answers will vary.
   Q2. Questions and answers will vary.
   Q3. Questions and answers will vary.
6. $36

Posttest Answer Key

1. $\frac{27}{3}$
2. $\frac{43}{8}$
3. $\frac{9}{4}$
4. 24
5. $18
6. $\frac{4}{9}$
7. $3\frac{3}{4}$ pounds
8. $1\frac{1}{2}$
9. $\frac{5}{9}$
10. 3
11. $9
12. 10 pounds
13. $\frac{9}{14}$
14. $7\frac{1}{2}$ hours
15. 21
16. $\frac{3}{5}$
17. $6\frac{2}{3}$
18. 15 centimeters
19. 10 boards
20. 9 feet

Posttest Evaluation Chart

If a student misses one or more problems in a skill area, assign a review of the practice pages for that skill.

Skill Area	Posttest Problem #	Skill Section	Review Page
Multiplication	1, 2, 8, 9, 15, 17	7–28	15, 29, 55, 56
Division	3, 4, 6, 10, 13, 16	37–53	54, 55, 56
Multiplication Word Problems	5, 11, 12, 20	30–35 57–68	36 69
Division Word Problems	7, 14, 18, 19	57–68	69
Life-Skills Math	All	70–73	74

RATIO AND PROPORTION

The concepts of ratio and proportion are key to the development of many topics in mathematics. Ratios compare sets of objects, things, or people or describe a rate. When a ratio is expressed as a rate, the quantities being compared have different names: miles and hours, miles and gallons, etc.

Two ratios that are equal to each other are called a proportion. When introducing proportions, it is useful to show students the similarities between proportions and the equivalent fractions that they have already studied.

Student Glossary

Acquainting students with definitions of key math terms and life-skills concepts will enhance their mastery of the materials. Below are words defined in the student text. A glossary is provided at the end of the student text and on page 243 of the *Teacher's Resource Guide and Answer Key.*

acre
average
cross product
denominator
ingredients
number relation symbol
order
per
proportion
ratio
simplify (reduce)
term
unit price
unit rate

STUDENT PAGE
7
Meaning of Ratio

Everyday Ratios and Rates

Discuss comparisons that students use in their everyday lives. Introduce a few comparisons and then challenge students to develop a list of their own.

EXAMPLES

Miles traveled to gallons of gasoline

Dollars earned each hour

One teacher for every 28 students

Number of hits for times at bat

A free car wash for every 10 tickets

Recent poll: 2 out of 3 students walk to school

Discuss the three ways of writing these comparisons in ratio form.

EXAMPLES

A 28 miles on 1 gallon of gasoline

28 to 1; 28:1; $\frac{28}{1}$

B Earned $9.35 each hour

9.35 to 1; 9.35:1; $\frac{9.35}{1}$

STUDENT PAGE
8
Meaning of Ratio

Newspaper Ratios

After introducing the idea of ratio, have students write ratios using everyday situations. Newspaper stories and advertisements express ratios in many ways. Ask students to bring in news articles that show relationships that can be converted to ratios. Have students express the ratios in three ways using a chart.

EXAMPLES

sports reports, price per pound, statistical information, etc.

2 pairs for $12		
2 to 12	2:12	$\frac{2}{12}$

Grapefruit juice: 3 cans for $1		
3 to 1	3:1	$\frac{3}{1}$

Explain two differences with writing ratios in fraction form:

1. Do not simplify, or reduce, the ratio as in the first example above.
2. Keep a denominator of 1 in a ratio as in the second example above.

STUDENT PAGE
10
Meaning of Ratio

Compare the Lengths

Have students write ratios that compare the lengths of line segments.

EXAMPLES

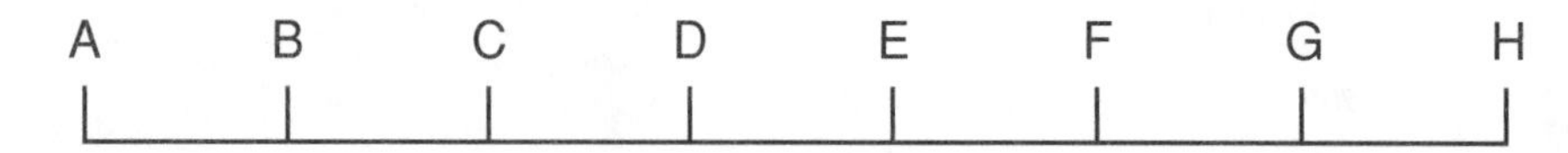

$\overline{AC}$ to $\overline{AD}$ (Answer: 2 to 3)

$\overline{AB}$ to $\overline{AH}$ (Answer: 1 to 7)

$\overline{HE}$ to $\overline{BG}$ (Answer: 3 to 5)

$\overline{FB}$ to $\overline{GD}$ (Answer: 4 to 3)

STUDENT PAGE
11
Meaning of Ratio

Using a Ruler

Draw a line segment 2 inches long. Draw another line segment so the ratio of the two lengths is 1:2. Have students draw their own ratios.

EXAMPLES

A Draw two line segments that have a ratio of 2:3.

B Draw two line segments that have a ratio of 1:4.

Discuss the measurements and compare students' results.

STUDENT PAGE
18
Ratio Applications

Find the Missing Term

For each table, have students complete equivalent ratios. Have students make their own tables to challenge the class.

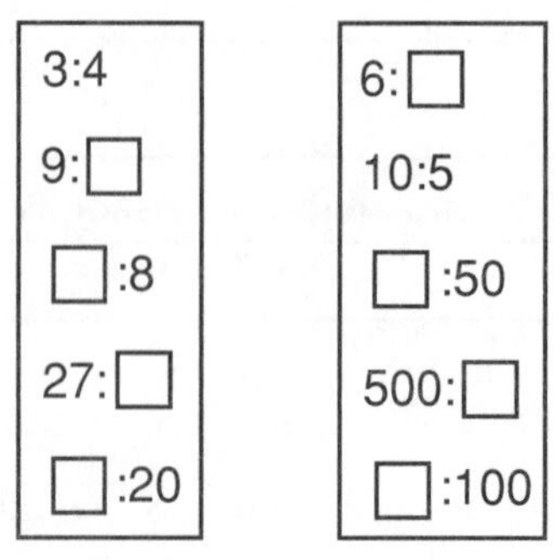

STUDENT PAGE
28
Ratio Applications

Measurement Ratios

Explain to students that they must express measurement ratios by using the same units. To do this, they must first convert the larger unit into smaller units. For example: to compare pennies to dollars, both must be in the form of pennies. Therefore, when comparing 17 pennies to 1 dollar, the ratio is 17:100. Have students develop questions involving conversions and quiz each other.

EXAMPLES

17 minutes to 1 hour (Answer: 17:60)

6 inches to 1 yard (Answer: 6:36)

1 dime to 3 quarters (Answer: 10:75)

1 pound to 12 ounces (Answer: 16:12)

3 items to a dozen (Answer: 3:12)

2 feet to 1 yard (Answer: 2:3)

STUDENT PAGE
35
Ratio Problem Solving

Ratio Applications

On the board, write ratio problems that students can relate to. Have students solve the problems.

EXAMPLE

If you buy 2 pounds of oranges for $1.29, how much will 4 pounds of oranges cost?

After students solve the problem, substitute different numbers. This provides more interesting practice than simply assigning a set of ratio problems from a textbook or worksheet.

Using newspapers and store advertisements, have students develop their own questions for classmates to solve.

STUDENT PAGE
43
Meaning of Proportion

Set Up a Table

Set up a table to show equal ratios.

Triangles	Circles
△△	○○○
2	3
4	6
6	9
8	12
10	15

Equal ratios $\frac{2}{3} = \frac{4}{6}$ $\frac{2}{3} = \frac{6}{9}$ $\frac{2}{3} = \frac{8}{12}$, etc.

Establish a pattern with the students. They can see that the ratio is always $\frac{2}{3}$ and that different proportions can be set up based on this relationship.

Make up several tables using different ratios. Let students make their own tables and discuss the tables with the class. Have the class review and comment on the different tables.

STUDENT PAGE
51
Meaning of Proportion

Mentally Figure Equal Ratios

Have students answer oral or written questions similar to the following.

EXAMPLES

2 is to 3 as what is to 6? (Answer: 4)

What is to 6 as 2 is to 18? (Answer: 1)

1 is to 4 as what is to 24? (Answer: 6)

1 is to 8 as 4 is to what? (Answer: 32)

Adjust the level of difficulty to meet the abilities of your students.

STUDENT PAGE
61
Proportion Applications

Unit Prices

Go over the proportions in each example below and have students find the unit price. Then assign students to find five examples in the newspaper, at the grocery store, or at the pharmacy. Have them find the unit price in their own examples.

EXAMPLES

3 pounds of apples for $2.97 (Unit price: $.99)

1 dozen doughnuts for $4.32 (Unit price: $.36)

2 pairs of shorts for $6.98 (Unit price: $3.49)

6 grapefruit for $3.30 (Unit price: $.55)

3 bars of soap for $1.98 (Unit price: $.66)

STUDENT PAGE
62
Proportion Applications

Find the Cost

Let students find the unit cost of different types of fruit and have them work with the information.

EXAMPLES

Bananas	3 for $.45
Apples	2 for $.50
Oranges	4 for $.48

List the fruits you can buy for exactly $.77.
(Answer: 1 banana, 2 apples, and 1 orange)

List the fruits you can buy for exactly $1.04.
(Answer: 2 bananas, 2 apples, and 2 oranges)

List the fruits you can buy for exactly $1.50.
(Answer: 6 apples or 5 oranges and 6 bananas)

STUDENT PAGE
63
Proportion Applications

Compare the Prices

Students can learn to use proportion to figure out the most economical purchases. Have them use advertisements for products that are sold in two sizes. They can determine which size is more economical by setting up ratios of *size to purchase price* and finding which is the better buy.

Many supermarkets advertise specials of the week.

1. Check store ads for the same item in different store fliers or advertisements.
2. Figure the unit cost of each item.
3. Determine which store gives the best buy.

STUDENT PAGE
65
Proportion Problem Solving

Recipes and Ratios

Write a recipe on the board and have students write how much of each ingredient should be used for different numbers of servings. Have them set up and solve a proportion for each serving size.

EXAMPLE

Scalloped Potatoes —Serves 6
6 medium potatoes
2 tablespoons flour
4 tablespoons butter

How much of each ingredient would be used for:

3 servings: $\frac{3 \text{ servings}}{6 \text{ servings}} = \frac{n \text{ potatoes}}{6 \text{ potatoes}}$

12 servings

9 servings

2 servings

4 servings

STUDENT PAGE
66
Proportion
Problem
Solving

Use a Calculator

Since calculators can be used effectively for proportion and percent applications, teachers should stress the use of calculators to solve proportions. Have students set up proportions from a list of problems and solve them using a calculator. Students can focus on learning to use proportions to solve a variety of problems without getting caught up in tedious calculations. Instruction on calculator use should also include work on estimation and ways of checking if the results are reasonable.

EXAMPLE

$$\frac{n}{120} = \frac{30}{40}$$

$n \times 40 = 30 \times 120$ ← use calculator to simplify multiplication

$n \times 40 = 3600$

$n = 3600 \div 40$ ← use calculator to simplify division

$n = 90$

NOTE: Students who are preparing to take the GED Test should become familiar with the Casio *fx-260SOLAR* calculator. This is the calculator that students use on Part 1 of the Mathematics Test.

STUDENT PAGE

69
Proportion
Problem
Solving

Using Maps

Set up workstations and place different state maps at each one. Make up questions on worksheets for each station. Have students move in teams of two through all workstations.

EXAMPLE QUESTIONS

What is the scale used on the map?

Choose two cities and find the distance between them.

Name two cities that are about 100 miles apart.

The length and width of the state equal about how many miles?

STUDENT PAGE
70
Proportion Problem Solving

Proportion Application

On the board, write proportion problems that have meaning to your students. Have students make a chart for each problem.

EXAMPLE

At the pizza parlor, 15 pizzas are made every 20 minutes. At the same rate, how many pizzas could be made in 60 minutes?

Pizzas	15	n
Minutes	20	60

Pretest

1. 3:5
2. 2:3
3. 7 to 1
4. 1:3 or 1 to 3
5. 3:1 or 3 to 1
6. 5:8 or 5 to 8
7. 3:4 or 3 to 4
8. 4:3 or 4 to 3
9. $10.50
10. 24 miles per gallon
11. $m = 30$
12. $x = 9$
13. $c = 25$
14. $n = 7\frac{1}{5}$
15. $y = 8\frac{3}{4}$
16. 9 tablespoons
17. $2\frac{1}{2}$ inches or 2.5 inches
18. 16 hits
19. $49.50
20. $84

Pretest Evaluation Chart

If a student misses any problems in a skill area, assign the practice pages for that skill. However, students may need to complete all practice pages to reinforce areas of weakness.

Skill Area	Pretest Problem #	Skill Section	Review Page
Meaning of Ratio	1, 2, 3	7–16	17
Ratio Applications	4, 5, 6, 7	18–32	33
Ratio Problem Solving	8, 9, 10	34–41	42
Meaning of Proportion	11, 12, 13, 14, 15	43–55	56
Proportion Applications	All	57–63	64
Proportion Applications & Problem Solving	16, 17, 18, 19, 20	65–73	74

Page 7: What Is Ratio?

1. a) 5 to 3 b) 5:3 c) $\frac{5}{3}$
2. a) 3 to 8 b) 3:8 c) $\frac{3}{8}$
3. a) 5 to 8 b) 5:8 c) $\frac{5}{8}$

Page 8: Write the Ratios

1. a) 2 to 3 b) 2:3 c) $\frac{2}{3}$
2. a) 3 to 2 b) 3:2 c) $\frac{3}{2}$
3. a) 2 to 5 b) 2:5 c) $\frac{2}{5}$
4. a) 3 to 5 b) 3:5 c) $\frac{3}{5}$
5. a) 5 to 2 b) 5:2 c) $\frac{5}{2}$

Page 9: Ratios as Fractions

1. $\frac{1}{5}$
2. $\frac{2}{6}$
3. $\frac{3}{4}$
4. $\frac{1}{3}$
5. $\frac{5}{7}$
6. $\frac{3}{5}$

Page 10: Compare the Shapes

1. $4:7 = \frac{4}{7}$
2. $7:4 = \frac{7}{4}$
3. $4:11 = \frac{4}{11}$
4. $7:11 = \frac{7}{11}$
5. $11:4 = \frac{11}{4}$
6. $11:7 = \frac{11}{7}$
7. $5:7 = \frac{5}{7}$
8. $7:5 = \frac{7}{5}$

Page 11: Draw the Ratios

1. 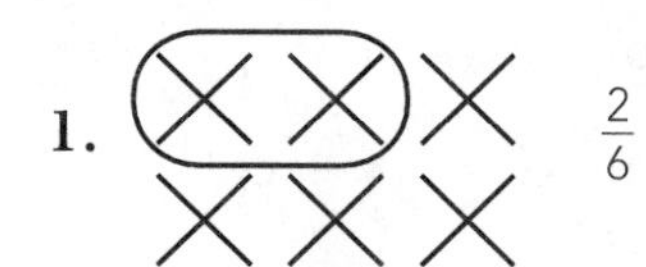 $\frac{2}{6}$

2. $\frac{6}{8}$

3. 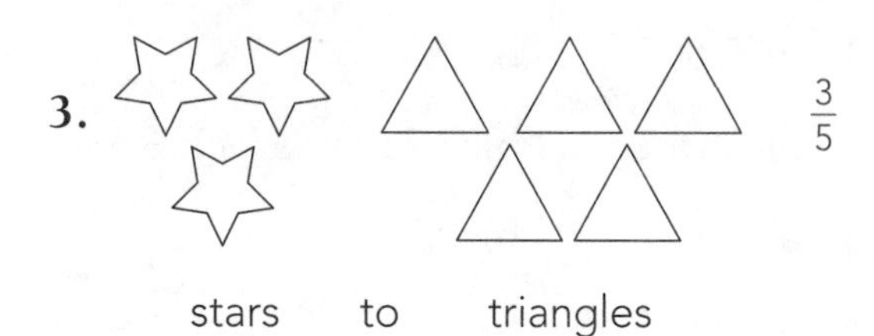 $\frac{3}{5}$

stars to triangles

4. 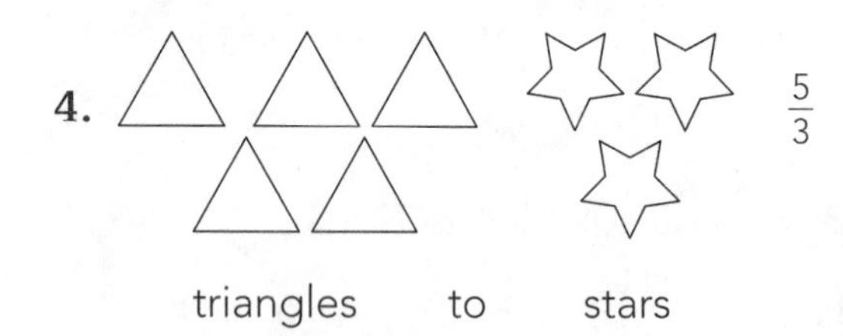 $\frac{5}{3}$

triangles to stars

Page 12: Write the Ratios as Fractions

1. $\frac{17}{30}$
2. $\frac{2}{3}$
3. $\frac{10}{2}$
4. $\frac{406}{7}$
5. $\frac{5}{12}$
6. $\frac{120}{8}$
7. $\frac{3}{36}$
8. $\frac{6}{10}$

Page 13: Simplify the Ratios

1. $\frac{5}{20} = \frac{1}{4}$
2. $\frac{12}{15} = \frac{4}{5}$
3. $\frac{15}{30} = \frac{1}{2}$
4. $\frac{4}{12} = \frac{1}{3}$
5. $\frac{8}{64} = \frac{1}{8}$
6. $\frac{6}{8} = \frac{3}{4}$
7. $\frac{8}{24} = \frac{1}{3}$
8. $\frac{16}{18} = \frac{8}{9}$
9. $\frac{9}{36} = \frac{1}{4}$
10. $\frac{21}{24} = \frac{7}{8}$
11. $\frac{7}{49} = \frac{1}{7}$
12. $\frac{6}{18} = \frac{1}{3}$
13. $\frac{28}{35} = \frac{4}{5}$
14. $\frac{10}{25} = \frac{2}{5}$
15. $\frac{11}{22} = \frac{1}{2}$
16. $\frac{5}{20} = \frac{1}{4}$

Page 14: Denominator of 1

1. $\frac{8}{2} = \frac{4}{1}$
2. $\frac{9}{3} = \frac{3}{1}$
3. $\frac{15}{5} = \frac{3}{1}$
4. $\frac{36}{6} = \frac{6}{1}$
5. $\frac{50}{10} = \frac{5}{1}$
6. $\frac{24}{4} = \frac{6}{1}$
7. $\frac{60}{10} = \frac{6}{1}$
8. $\frac{56}{8} = \frac{7}{1}$
9. $\frac{12}{4} = \frac{3}{1}$
10. $\frac{63}{9} = \frac{7}{1}$
11. $\frac{70}{7} = \frac{10}{1}$
12. $\frac{64}{8} = \frac{8}{1}$
13. $\frac{49}{7} = \frac{7}{1}$
14. $\frac{100}{5} = \frac{20}{1}$

Page 15: Equal Ratios Are Equal Fractions

1. $\frac{1}{3} = \frac{2}{6} = \frac{3}{9} = \frac{4}{12} = \frac{5}{15} = \frac{6}{18} = \frac{7}{21} = \frac{8}{24} = \frac{9}{27}$
2. $\frac{5}{1} = \frac{10}{2} = \frac{15}{3} = \frac{20}{4} = \frac{25}{5} = \frac{30}{6} = \frac{35}{7} = \frac{40}{8} = \frac{45}{9}$
3. $\underbrace{\frac{3}{1} = \frac{6}{2} = \frac{9}{3}} = \frac{12}{4} = \frac{15}{5} = \frac{18}{6} = \frac{21}{7} = \frac{24}{8} = \frac{27}{9}$

   $\frac{3}{1} \times \frac{3}{3} = \frac{9}{3}$
4. $\frac{1}{5} = \frac{2}{10} = \frac{3}{15} = \frac{4}{20} = \frac{5}{25} = \frac{6}{30} = \frac{7}{35} = \frac{8}{40} = \frac{9}{45}$

Page 16: Equivalent Ratios

1. $\frac{1}{2} = \frac{4}{8}$
2. $\frac{5}{6} = \frac{15}{18}$
3. $\frac{7}{8} = \frac{14}{16}$
4. $\frac{1}{5} = \frac{9}{45}$
5. $\frac{6}{7} = \frac{18}{21}$
6. $\frac{3}{4} = \frac{15}{20}$
7. $\frac{2}{9} = \frac{8}{36}$
8. $\frac{4}{5} = \frac{20}{25}$
9. $\frac{5}{20} = \frac{1}{4}$
10. $\frac{18}{45} = \frac{2}{5}$
11. $\frac{2}{14} = \frac{1}{7}$
12. $\frac{21}{49} = \frac{3}{7}$
13. $\frac{9}{24} = \frac{3}{8}$
14. $\frac{5}{6} = \frac{10}{12}$
15. $\frac{3}{4} = \frac{27}{36}$
16. $\frac{5}{3} = \frac{10}{6}$
17. $\frac{5}{4} = \frac{20}{16}$
18. $\frac{40}{64} = \frac{5}{8}$
19. $\frac{5}{7} = \frac{30}{42}$
20. $\frac{3}{7} = \frac{21}{49}$
21. $\frac{10}{25} = \frac{2}{5}$
22. $\frac{1}{8} = \frac{5}{40}$
23. $\frac{3}{27} = \frac{1}{9}$
24. $\frac{3}{2} = \frac{6}{4}$

Page 17: Meaning of Ratio Review

1. $\frac{5}{15} = \frac{1}{3}$
2. $\frac{15}{12} = \frac{5}{4}$
3. $\frac{3}{10}$
4. $\frac{3}{5}$
5. $\frac{3}{6} = \frac{1}{2}$
6. 3:4
7. 1:4
8. 2:3
9. 1:4
10. 1:5
11. $\frac{1}{2}$
12. $\frac{1}{6}$
13. $\frac{1}{10}$
14. $\frac{1}{10}$
15. $\frac{1}{4}$
16. $\frac{3}{4} = \frac{15}{20}$
17. $\frac{18}{30} = \frac{3}{5}$
18. $\frac{3}{7} = \frac{18}{42}$
19. $\frac{5}{3} = \frac{15}{9}$
20. $\frac{5}{6} = \frac{20}{24}$

Page 18: Find a Pattern

1.

Cans	1	2	3	4	5	6	7	8
Tennis Balls	3	6	9	12	15	18	21	24

2.

Packs of Gum	1	3	5	7	9	11	13	15
Sticks of Gum	5	15	25	35	45	55	65	75

Think: $\frac{1}{5} \times \frac{9}{9} = \frac{9}{45}$

3.

Touchdowns	1	7	3	2	9	4	6	5
Points	6	42	18	12	54	24	36	30

Think: $\frac{1}{6} \times \frac{2}{2} = \frac{2}{12}$

4.

Dollars	1	3	7	2	10	6	9	4
Quarters	4	12	28	8	40	24	36	16

Think: $\frac{1}{4} \times \frac{3}{3} = \frac{3}{12}$

Page 19: Fill in the Table

1.

Hours Worked	1	2	3	8	40
Paid	\$9	\$18	\$27	\$72	\$360

Think: $\frac{1}{9} \times \frac{3}{3} = \frac{3}{27}$

2.

Hours Worked	1	2	3	4	5
Paid	\$8.75	\$17.50	\$26.25	\$35	\$43.75

Think: $\frac{1}{8.75} \times \frac{4}{4} = \frac{4}{35}$

3.

Hours Worked	1	8	10	40	80
Paid	\$6.25	\$50	\$62.50	\$250	\$500

4.

Hours Worked	1	4	8	20	40
Paid	\$12.95	\$51.80	\$103.60	\$259	\$518

Page 20: From Words to Ratios

1. $\frac{8}{2} = \frac{4}{1}$
2. $\frac{3}{12} = \frac{1}{4}$
3. $\frac{24}{3} = \frac{8}{1}$
4. $\frac{4}{72} = \frac{1}{18}$
5. $\frac{100}{5} = \frac{20}{1}$
6. $\frac{6}{10} = \frac{3}{5}$
7. $\frac{5}{35} = \frac{1}{7}$
8. $\frac{27}{3} = \frac{9}{1}$

Page 21: Unit Rates

1. $\frac{50}{1}$ 50 miles (per) hour
2. $\frac{3}{1}$ 3 tennis balls (per) can
3. $\frac{9}{1}$ 9 revolutions (each) minute
4. $\frac{75}{1}$ 75 tablets in (one) bottle
5. $\frac{25}{1}$ 25 miles (per) gallon

Page 22: Write as a Ratio in Fraction Form

1. $\frac{46}{1}$ (per)
2. $\frac{48}{5}$
3. $\frac{7}{20}$
4. $\frac{11}{9}$
5. $\frac{1}{5}$
6. $\frac{7}{1}$ (per)
7. $\frac{1}{28}$
8. $\frac{8}{9.85}$
9. $\frac{32}{1}$ (per)
10. $\frac{9}{11}$
11. $\frac{25}{1}$ (each)
12. $\frac{55}{1}$ (per)
13. $\frac{5}{16}$
14. $\frac{36}{1}$ (per)
15. $\frac{2}{1}$ (per)
16. $\frac{129}{8}$
17. $\frac{9}{1}$ (per)
18. $\frac{5}{1}$ (per)
19. $\frac{17}{1}$ (per)
20. $\frac{25}{3}$

Page 23: Writing Unit Rates

1. $\frac{85}{5} = \frac{17}{1}$ 17 miles per gallon
2. $\frac{165}{3} = \frac{55}{1}$ 55 miles per hour
3. $\frac{64}{4} = \frac{16}{1}$ 16 ounces per can
4. $\frac{56}{7} = \frac{8}{1}$ 8 apples per box
5. $\frac{63}{9} = \frac{7}{1}$ 7 ounces per cup
6. $\frac{21}{3} = \frac{7}{1}$ $7 per ticket
7. $\frac{360}{8} = \frac{45}{1}$ 45 miles per hour
8. $\frac{120}{5} = \frac{24}{1}$ 24 miles per gallon

Page 24: Comparing Unit Rates

1. $\frac{133}{7} = \frac{19}{1}$ $\frac{108}{6} = \frac{18}{1}$
   19 miles per hour (>) 18 miles per hour
2. $\frac{72}{3} = \frac{24}{1}$ $\frac{120}{5} = \frac{24}{1}$
   24 miles per gallon (=) 24 miles per gallon
3. $\frac{60}{4} = \frac{15}{1}$ $\frac{90}{5} = \frac{18}{1}$
   $15 for one ticket (<) $18 for one ticket
4. $\frac{425}{5} = \frac{85}{1}$ $\frac{294}{3} = \frac{98}{1}$
   85 meters per hour (<) 98 meters per hour

Page 25: Find the Rates

1. a) $24.00
   b) $64.00
2. a) 360 miles
   b) 180 miles
3. a) 345 miles
   b) 506 miles
4. a) $4.17
   b) $6.95
5. a) $24.75
   b) $57.75
6. a) $20.00
   b) $45.00

Page 26: Find the Cost

1. a) $1.09
   b) $2.18
   c) $3.27
   d) $4.36
2. a) $1.45
   b) $5.80
   c) $8.70
   d) $14.50
3. a) $1.73
   b) $3.46
   c) $19.03
   d) $25.95
4. a) $1.42
   b) $2.84
   c) $5.68
   d) $7.10
5. a) $3.09
   b) $6.18
   c) $15.45
   d) $21.63
6. a) $3.50
   b) $10.50
   c) $14.00
   d) $21.00

Page 27: Ratios as Rates

1. a) $24.00
   b) $36.00
   c) $48.00
2. a) $18.00
   b) $27.00
   c) $36.00
3. a) $1.64
   b) $2.46
   c) $3.28
4. a) $9.00
   b) $13.50
   c) $18.00
5. a) $30.00
   b) $50.00
   c) $70.00
6. a) $4.05
   b) $6.75
   c) $10.80
7. a) $14.85
   b) $24.75
   c) $39.60
8. a) $12.75
   b) $21.25
   c) $38.25

Page 28: Measurement Ratios

1. $\frac{3}{12} = \frac{1}{4}$
2. $\frac{12}{12} = \frac{1}{1}$
3. $\frac{36}{12} = \frac{3}{1}$
4. $\frac{6}{3} = \frac{2}{1}$
5. $\frac{3}{3} = \frac{1}{1}$
6. $\frac{1}{3} = \frac{1}{3}$

Page 29: Comparing Inches and Feet

1. $\frac{4}{24} = \frac{1}{6}$
2. $\frac{3}{12} = \frac{1}{4}$
3. $\frac{10}{60} = \frac{1}{6}$
4. $\frac{12}{36} = \frac{1}{3}$
5. $\frac{9}{12} = \frac{3}{4}$

Page 30: Comparing Feet and Yards

1. $\frac{4}{12} = \frac{1}{3}$
2. $\frac{3}{6} = \frac{1}{2}$
3. $\frac{6}{15} = \frac{2}{5}$
4. $\frac{5}{30} = \frac{1}{6}$
5. $\frac{2}{18} = \frac{1}{9}$

Page 31: Comparing Ounces and Pounds

1. $\frac{4}{16} = \frac{1}{4}$
2. $\frac{6}{48} = \frac{1}{8}$
3. $\frac{8}{32} = \frac{1}{4}$
4. $\frac{2}{32} = \frac{1}{16}$
5. $\frac{10}{80} = \frac{1}{8}$

Page 32: Money and Time Ratios

1. $\frac{6}{20} = \frac{3}{10}$
2. $\frac{2}{20} = \frac{1}{10}$
3. $\frac{2}{8} = \frac{1}{4}$
4. $\frac{15}{60} = \frac{1}{4}$
5. $\frac{60}{100} = \frac{3}{5}$
6. $\frac{50}{120} = \frac{5}{12}$

Page 33: Ratio Applications Review

1. a) 4 to 3
   b) 4:3
   c) $\frac{4}{3}$
2. $\frac{13}{26} = \frac{1}{2}$
3. $\frac{48}{6} = \frac{8}{1}$
4. $\frac{12}{32}$
5. $\frac{8}{12} = \frac{2}{3}$
6. $\frac{45}{3} = \frac{15}{1}$
7. $\frac{25}{1} = \frac{5}{1}$ $\frac{49}{7} = \frac{7}{1}$

   $5 for 1 ticket (<) $7 for 1 ticket
8. a) $2.15
   b) $3.44
   c) $4.30
9. $\frac{15}{24} = \frac{5}{8}$
10. $\frac{6}{12} = \frac{1}{2}$
11. $\frac{12}{16} = \frac{3}{4}$
12. $\frac{10}{40} = \frac{1}{4}$
13. $\frac{5}{60} = \frac{1}{12}$

Page 34: Ratio Applications

1. 60:130 = $\frac{60}{130} = \frac{6}{13}$
2. 130:20 = $\frac{130}{20} = \frac{13}{2}$
3. 50:130 = $\frac{50}{130} = \frac{5}{13}$
4. 130:60 = $\frac{130}{60} = \frac{13}{6}$
5. 110:130 = $\frac{110}{130} = \frac{11}{13}$
6. 5:12 = $\frac{5}{12}$
7. 4:12 = $\frac{4}{12} = \frac{1}{3}$
8. 12:4 = $\frac{12}{4} = \frac{3}{1}$

Page 35: Using Ratios

1. a) $9.96
   b) $7.47
2. a) $14.40
   b) $19.20
3. a) $45.00
   b) $37.50
4. a) $40.00
   b) $15.00

Page 36: Real-Life Ratios

1. a) $25.30
   b) $3.49
   c) $14.44
2. a) $86.64
   b) $59.85
   c) $27.92
3. $5.32
4. a) $13.96
   b) $43.32
   c) $101.20
5. a) $59.85
   b) $10.47
   c) $50.60
6. $5.56

Page 37: Real-Life Practice

1. a) $7.49
   b) $3.75
   c) $18.33
   d) $.19
2. a) $1.33
   b) $29.96
   c) $18.75
   d) $54.99
3. $5
4. a) $37.45
   b) $7.50
   c) $73.32
   d) $2.28
5. a) $109.98
   b) $30.00
   c) $6.84
   d) $67.41
6. $3.39

Page 38: Seeing Ratios in Word Problems

1. d) 15:5
2. d) 1:3
3. d) 74:121
4. b) 18:100

Page 39: Word Problem Practice

1. 1,800:300
2. 6:31
3. 9:21
4. 7:4
5. 5:8
6. 15:14

Page 40: Ratio Relationships

1. 3:5
2. 7:9
3. .90
4. 3:4
5. 9:3 = 3:1
6. a) 15:20 = 3:4
   b) 5:20 = 1:4
7. 20 miles
8. $63.60

Page 41: More Ratio Relationships

1. 3:8
2. 8:6
3. 5:14
4. 2:14
5. 5:4
6. 450:500 = 9:10
7. 60 miles
8. $2,250.00

Page 42: Ratio Problem-Solving Review

1. 5:6
2. 525 miles
3. 4:8
4. 156:98 = 78:49
5. 145:750 = 29:150
6. $39.95
7. 34:45
8. 725:2,450 = 29:98

Page 43: What Is a Proportion?

1. $\frac{3}{4} = \frac{6}{8}$
2. $\frac{15}{5} = \frac{3}{1}$
3. $\frac{1}{4} = \frac{3}{12}$
4. $\frac{3}{2} = \frac{15}{10}$
5. $\frac{4}{7} = \frac{16}{28}$
6. $\frac{5}{6} = \frac{10}{12}$
7. $\frac{3}{4} = \frac{15}{20}$
8. $\frac{5}{10} = \frac{1}{2}$
9. $\frac{6}{27} = \frac{2}{9}$
10. $\frac{40}{100} = \frac{4}{10}$
11. $\frac{9}{7} = \frac{18}{14}$
12. $\frac{75}{100} = \frac{3}{4}$

Page 44: Simplify One Ratio

1. $\frac{4}{8} = \frac{5}{10}$
2. $\frac{6}{15} = \frac{8}{20}$
3. $\frac{3}{18} = \frac{5}{30}$
4. $\frac{8}{12} = \frac{10}{15}$
5. $\frac{9}{12} = \frac{6}{8}$
6. $\frac{18}{24} = \frac{15}{20}$
7. $\frac{18}{3} = \frac{30}{5}$
8. $\frac{15}{9} = \frac{20}{12}$
9. $\frac{7}{21} = \frac{9}{27}$
10. $\frac{10}{14} = \frac{15}{21}$
11. $\frac{6}{2} = \frac{15}{5}$
12. $\frac{30}{25} = \frac{18}{15}$

Page 45: Read the Proportion

1. $\frac{2}{3} = \frac{6}{9}$
2. $\frac{1}{4} = \frac{2}{8}$
3. $\frac{7}{3} = \frac{14}{6}$
4. $\frac{2}{5} = \frac{4}{10}$
5. $\frac{5}{8} = \frac{10}{16}$
6. $\frac{8}{12} = \frac{4}{6}$
7. $\frac{50}{25} = \frac{2}{1}$

Page 46: Two Equal Ratios

1. $\frac{2}{3} = \frac{6}{9}$
   $2 \times 9 = 6 \times 3$
   $18 = 18$
2. $\frac{7}{3} = \frac{21}{9}$
   $7 \times 9 = 21 \times 3$
   $63 = 63$
3. $\frac{4}{5} = \frac{16}{20}$
   $4 \times 20 = 16 \times 5$
   $80 = 80$
4. $\frac{8}{5} = \frac{40}{25}$
   $8 \times 25 = 40 \times 25$
   $200 = 200$
5. $\frac{6}{18} = \frac{12}{36}$
   $6 \times 36 = 12 \times 18$
   $216 = 216$
6. $\frac{5}{1} = \frac{10}{2}$
   $5 \times 2 = 10 \times 1$
   $10 = 10$
7. $\frac{8}{3} = \frac{16}{6}$
   $8 \times 6 = 16 \times 3$
   $48 = 48$
8. $\frac{7}{8} = \frac{91}{104}$
   $7 \times 104 = 91 \times 8$
   $728 = 728$
9. $\frac{10}{15} = \frac{30}{45}$
   $10 \times 45 = 30 \times 15$
   $450 = 450$

Page 47: Cross Products

1. $\frac{1}{4} \times \frac{3}{12}$

$1 \times 12 \quad 3 \times 4$
$12 = 12$

2. $\frac{5}{6} \times \frac{2}{3}$

$5 \times 3 \quad 2 \times 6$
$15 \neq 12$

3. $\frac{21}{28} \times \frac{3}{4}$

$21 \times 4 \quad 3 \times 28$
$84 = 84$

4. $\frac{42}{56} \times \frac{7}{9}$

$42 \times 9 \quad 7 \times 56$
$378 \neq 392$

5. $\frac{9}{13} \times \frac{63}{91}$

$9 \times 91 \quad 63 \times 13$
$819 = 819$

6. $\frac{7}{3} \times \frac{14}{6}$

$7 \times 6 \quad 14 \times 3$
$42 = 42$

7. $\frac{5}{12} \times \frac{80}{192}$

$5 \times 192 \quad 80 \times 12$
$960 = 960$

8. $\frac{4}{7} \times \frac{52}{91}$

$4 \times 91 \quad 52 \times 7$
$364 = 364$

9. $\frac{12}{15} \times \frac{24}{40}$

$12 \times 40 \quad 24 \times 15$
$480 \neq 360$

Page 48: Proportion Readiness

1. $10 = 10$
2. $9 \neq 10$
3. $72 = 72$
4. $432 \neq 162$
5. $96 \neq 72$
6. $108 = 108$
7. $128 = 128$
8. $132 \neq 104$
9. $72 = 72$
10. $360 = 360$
11. $126 = 126$
12. $105 \neq 108$

Page 49: Find the Unknown Term

1. 24
2. 20
3. 2
4. 3
5. 6
6. 12

Page 50: Solve and Check

1. 20
2. 10
3. 35
4. 9
5. 8
6. 15
7. 90
8. 3
9. 4

Page 51: Apply Your Skills

1. 5
2. 12
3. 15
4. 6
5. 49
6. 45
7. 15
8. 40
9. 1,000
10. 28
11. 9
12. 3
13. 32
14. 10
15. 18

Page 52: Proportions with Fractions

1. $3\frac{1}{3}$
2. $4\frac{4}{5}$
3. $11\frac{2}{3}$
4. $10\frac{1}{2}$
5. $7\frac{7}{8}$
6. $9\frac{1}{7}$
7. $3\frac{3}{5}$
8. $15\frac{3}{4}$

Page 53: Proportions with Decimals

1. 4.8
2. 2
3. 13.09
4. 2.36
5. 35
6. 6
7. 1.6
8. 12

Page 54: Missing Term

1. $\frac{3}{4} = \frac{n}{16}$ $\quad n = 12$
2. $\frac{4}{6} = \frac{10}{n}$ $\quad n = 15$
3. $\frac{n}{20} = \frac{5}{8}$ $\quad n = 12\frac{1}{2}$
4. $\frac{2.5}{n} = \frac{15}{18}$ $\quad n = 3$
5. $\frac{2.78}{2} = \frac{n}{5}$ $\quad n = 6.95$
6. $\frac{8}{5} = \frac{13}{n}$ $\quad n = 8\frac{1}{8}$
7. $\frac{3.25}{1} = \frac{n}{6}$ $\quad n = 19.5$
8. $\frac{n}{4} = \frac{15}{6}$ $\quad n = 10$

Page 55: More Missing Terms

1. $\frac{2}{5} = \frac{n}{15}$ $n = 6$
2. $\frac{n}{35} = \frac{1}{5}$ $n = 7$
3. $\frac{n}{4} = \frac{6.4}{3.2}$ $n = 8$
4. $\frac{7}{4} = \frac{10.5}{n}$ $n = 6$
5. $\frac{5}{8} = \frac{n}{32}$ $n = 20$
6. $\frac{7.45}{1} = \frac{n}{5}$ $n = 37.25$
7. $\frac{n}{20} = \frac{3}{8}$ $n = 7.5$
8. $\frac{n}{5} = \frac{12}{6}$ $n = 10$

Page 56: Meaning of Proportion Review

1. $\frac{5}{6} = \frac{15}{18}$
2. $\frac{4}{8} = \frac{6}{12}$
3. $\frac{9}{27} = \frac{6}{18}$
4. $7 \times 9 = 21 \times 3$
   $63 = 63$
5. $315 \neq 320$
6. $n = 60 \div 5$
   $n = 12$
7. $4 \times n = 60$
   $n = 15$
8. $54 = 5 \times n$
   $n = 10\frac{4}{5}$
9. $n \times 2.49 = 12.45$
   $n = 5$
10. $\frac{7.25}{1} = \frac{n}{6}$
    $n = 43.5$
11. $\frac{n}{4} = \frac{15}{10}$
    $n \times 10 = 60$
    $n = 6$

Page 57: Proportions in Problem Solving

1. cans to cost $\frac{5}{\$6.50}$ $\frac{\text{cans}}{\text{cost}}$
2. cans to cost $\frac{9}{n}$ $\frac{\text{cans}}{\text{cost}}$
3. $\frac{5}{\$6.50}\ \frac{\text{cans}}{\text{cost}} = \frac{9}{n}\ \frac{\text{cans}}{\text{cost}}$

Page 58: Setting Up Proportions

1. $\frac{3}{24}\ \frac{\text{miles}}{\text{minutes}} = \frac{10}{n}\ \frac{\text{miles}}{\text{minutes}}$
2. $\frac{675}{n}\ \frac{\text{miles}}{\text{hours}} = \frac{45}{1}\ \frac{\text{miles}}{\text{hour}}$
3. $\frac{2}{7}\ \frac{\text{hits}}{\text{at bat}} = \frac{n}{56}\ \frac{\text{hits}}{\text{at bat}}$
4. $\frac{5}{\$3.95}\ \frac{\text{pounds}}{\text{cost}} = \frac{13}{n}\ \frac{\text{pounds}}{\text{cost}}$

Page 59: Check Your Proportions

1. $\frac{3}{\$5.94}\ \frac{\text{yards}}{\text{cost}} = \frac{2}{\$3.96}\ \frac{\text{yards}}{\text{cost}}$

   Check: $3 \times \$5.94 = 2 \times \$3.96$
   $\$11.88 = \$11.88$

2. $\frac{255}{3}\ \frac{\text{bushels}}{\text{acres}} = \frac{1{,}700}{20}\ \frac{\text{bushels}}{\text{acres}}$

   Check: $255 \times 20 = 1{,}700 \times 3$
   $5{,}100 = 5{,}100$

3. $\frac{3}{900}\ \frac{\text{pounds}}{\text{square feet}} = \frac{7}{2{,}100}\ \frac{\text{pounds}}{\text{square feet}}$

   Check: $3 \times 2{,}100 = 7 \times 900$
   $6{,}300 = 6{,}300$

4. $\frac{120}{3}\ \frac{\text{miles}}{\text{days}} = \frac{280}{7}\ \frac{\text{miles}}{\text{days}}$

   Check: $120 \times 7 = 280 \times 3$
   $840 = 840$

Page 60: Using Proportions

1. $3 \times 56 = n \times 2$
   $168 = n \times 2$
   $84 \text{ boys} = n$

2. $25 \times 6 = n \times 1$
   $150 = n \times 1$
   $150 \text{ minutes} = n$

3. $13 \times n = 1 \times 273$
   $13 \times n = 273$
   $n = 21 \text{ miles per gallon}$

4. $245 \times 7 = n \times 5$
   $1{,}715 = n \times 5$
   $\$343.00 \text{ saved in 7 weeks} = n$

Page 61: Unit Prices

1. $\frac{32 \text{ ounces}}{\$2.56 \text{ cost}} = \frac{1 \text{ ounce}}{n \text{ cost}}$
   The unit price is $.08 per ounce.

2. $\frac{6 \text{ ounces}}{\$1.32 \text{ cost}} = \frac{1 \text{ ounce}}{n \text{ cost}}$
   The unit price is $.22 per ounce.

3. $\frac{\$4.99 \text{ cost}}{50 \text{ tablets}} = \frac{n \text{ cost}}{1 \text{ tablet}}$
   The unit price is $.10 per tablet.

4. $\frac{\$2.94 \text{ cost}}{7 \text{ ounces}} = \frac{n \text{ cost}}{1 \text{ ounce}}$
   The unit price is $.42 per ounce.

Page 62: Find the Unit Costs

1. $\frac{6 \text{ pounds}}{7.20 \text{ dollars}} = \frac{1 \text{ pound}}{n \text{ dollars}}$
   Unit price: $1.20 per pound

2. $\frac{8 \text{ pounds}}{2.16 \text{ dollars}} = \frac{1 \text{ pound}}{n \text{ dollars}}$
   Unit price: $.27 per pound

3. $\frac{5 \text{ grapefruit}}{3.45 \text{ dollars}} = \frac{1 \text{ grapefruit}}{n \text{ dollars}}$
   Unit price: $.69 each

4. $\frac{3 \text{ pounds}}{4.50 \text{ dollars}} = \frac{1 \text{ pound}}{n \text{ dollars}}$
   Unit price: $1.50 per pound

5. $\frac{16 \text{ ounces}}{.79 \text{ dollars}} = \frac{1 \text{ ounce}}{n \text{ dollars}}$
   Unit price: $.05 per ounce

6. $\frac{7 \text{ apples}}{2.90 \text{ dollars}} = \frac{1 \text{ apple}}{n \text{ dollars}}$
   Unit price: $.41 each

Page 63: Comparison Shopping

1. $\frac{9 \text{ ounces}}{2.70 \text{ dollars}} = \frac{1 \text{ ounce}}{n \text{ dollars}}$
   The unit price is $.30 per ounce.
   $\frac{15 \text{ ounces}}{3.75 \text{ dollars}} = \frac{1 \text{ ounce}}{n \text{ dollars}}$
   The unit price is $.25 per ounce.
   The 15-ounce size is cheaper per ounce.

2. $\frac{6 \text{ bars}}{3.42 \text{ dollars}} = \frac{1 \text{ bar}}{n \text{ dollars}}$
   The unit price is $.57 per candy bar.
   $\frac{3 \text{ bars}}{1.56 \text{ dollars}} = \frac{1 \text{ bar}}{n \text{ dollars}}$
   The unit price is $.52 per candy bar.
   The package containing 3 candy bars is cheaper per bar.

3. $\frac{200 \text{ feet}}{6.00 \text{ dollars}} = \frac{1 \text{ foot}}{n \text{ dollars}}$

The unit price is $.03 per foot.

$\frac{25 \text{ feet}}{1.00 \text{ dollars}} = \frac{1 \text{ foot}}{n \text{ dollars}}$

The unit price is $.04 per foot.

The 200-foot size is cheaper per foot.

4. $\frac{16 \text{ ounces}}{2.56 \text{ dollars}} = \frac{1 \text{ ounce}}{n \text{ dollars}}$

The unit price is $.16 per ounce.

$\frac{6 \text{ ounces}}{1.32 \text{ dollars}} = \frac{1 \text{ ounce}}{n \text{ dollars}}$

The unit price is $.22 per ounce.

The 16-ounce size is cheaper per ounce.

Page 64: Proportion Applications Review

1. $\frac{330}{n} = \frac{55}{1}$
2. $\frac{3}{8} = \frac{n}{56}$
3. $\frac{235}{2} = \frac{n}{8}$
4. $\frac{8}{21.36} = \frac{12}{n}$
5. .13
6. .23
7. 1.03
8. .06

Page 65: Changing Recipes

A. $5 = n$

5 eggs are needed to make 25 waffles.

1. $\frac{2 \text{ eggs}}{24 \text{ cookies}} = \frac{n \text{ eggs}}{60 \text{ cookies}}$

5 eggs are needed to make 60 cookies.

2. $\frac{3 \text{ ounces}}{9 \text{ servings}} = \frac{4 \text{ ounces}}{n \text{ servings}}$

4 ounces of cream cheese will make 12 servings.

3. $\frac{12 \text{ people}}{4 \text{ oranges}} = \frac{18 \text{ people}}{n \text{ oranges}}$

6 oranges are needed to make punch for 18 people.

4. $\frac{3 \text{ teaspoons}}{36 \text{ servings}} = \frac{n \text{ teaspoons}}{12 \text{ servings}}$

1 teaspoon of butter is needed for 12 servings.

5. $\frac{6 \text{ tablespoons}}{4 \text{ servings}} = \frac{9 \text{ tablespoons}}{n \text{ servings}}$

9 tablespoons of milk will make 6 servings.

6. $\frac{6 \text{ teaspoons}}{30 \text{ biscuits}} = \frac{n \text{ teaspoons}}{40 \text{ biscuits}}$

8 teaspoons of baking powder are needed to make 40 biscuits.

Page 66: Figuring Costs

A. $n = \$19.10$

Five boxes of cereal cost $19.10.

1. $\frac{4.80 \text{ dollars}}{1 \text{ square yard}} = \frac{n \text{ dollars}}{8 \text{ square yards}}$

8 square yards will cost $38.40.

2. $\frac{2 \text{ notebooks}}{2.34 \text{ dollars}} = \frac{n \text{ notebooks}}{7.02 \text{ dollars}}$

Patricia could buy 6 notebooks for $7.02.

3. $\frac{3 \text{ candy bars}}{2.25 \text{ dollars}} = \frac{10 \text{ candy bars}}{n \text{ dollars}}$

10 candy bars would cost $7.50.

4. $\frac{6 \text{ pounds}}{1.50 \text{ dollars}} = \frac{13 \text{ pounds}}{n \text{ dollars}}$

13 pounds of potatoes will cost $3.25.

5. $\frac{3 \text{ cans}}{2.26 \text{ dollars}} = \frac{9 \text{ cans}}{n \text{ dollars}}$

Sherlene spent $6.78 for 9 cans of soup.

6. $\frac{1.89 \text{ dollars}}{3 \text{ oranges}} = \frac{n \text{ dollars}}{7 \text{ oranges}}$

7 oranges will cost $4.41.

Page 67: Travel Plans

A. $n = 7$ It will take Kevin 7 hours.

1. $\frac{21 \text{ miles}}{1 \text{ gallon}} = \frac{315 \text{ miles}}{n \text{ gallons}}$

15 gallons of gasoline will be used on a 315-mile trip.

2. $\frac{1{,}365 \text{ miles}}{3 \text{ hours}} = \frac{n \text{ miles}}{5 \text{ hours}}$

A jet can fly 2,275 miles in 5 hours.

3. $\frac{55 \text{ miles}}{1 \text{ hour}} = \frac{220 \text{ miles}}{n \text{ hours}}$

It would take the train 4 hours to travel 220 miles.

4. $\frac{45 \text{ miles}}{1 \text{ hour}} = \frac{n \text{ miles}}{4.5 \text{ hours}}$

The train will travel 202.5 miles in 4.5 hours.

5. $\frac{.40 \text{ dollars}}{25 \text{ miles}} = \frac{n \text{ dollars}}{175 \text{ miles}}$

It would cost $2.80 to travel 175 miles.

6. $\frac{104 \text{ miles}}{2 \text{ hours}} = \frac{260 \text{ miles}}{n \text{ hours}}$

It will take Benjamin 5 hours to travel 260 miles.

Page 68: Scale Drawings

1. $\frac{1 \text{ inch (map)}}{194 \text{ miles (actual)}} = \frac{2.5 \text{ inches (map)}}{n \text{ miles (actual)}}$

$n = 485$

The actual distance between El Paso and San Antonio is 485 miles.

2. $\frac{1 \text{ inch (map)}}{194 \text{ miles (actual)}} = \frac{1.5 \text{ inches (map)}}{n \text{ miles (actual)}}$

$n = 291$

The actual distance between Lubbock and Dallas is 291 miles.

3. $\frac{1 \text{ inch (map)}}{194 \text{ miles (actual)}} = \frac{2 \text{ inches (map)}}{n \text{ miles (actual)}}$

$n = 388$

The actual distance between Wichita Falls and Corpus Christi is 388 miles.

4. $\frac{1 \text{ inch (map)}}{194 \text{ miles (actual)}} = \frac{.5 \text{ inch (map)}}{n \text{ miles (actual)}}$

$n = 97$

The actual distance between Texarkana and Tyler is 97 miles.

Page 69: Map Applications

1. $\frac{1 \text{ inch}}{150 \text{ miles}} = \frac{4 \text{ inches}}{n \text{ miles}}$

The two cities are 600 miles apart.

2. $\frac{1 \text{ inch}}{45 \text{ miles}} = \frac{n \text{ inches}}{315 \text{ miles}}$

7 inches on the map represent 315 miles.

3. $\frac{1 \text{ inch}}{75 \text{ miles}} = \frac{3 \text{ inches}}{n \text{ miles}}$

3 inches on the map represent 225 miles.

4. $\frac{2 \text{ inches}}{150 \text{ miles}} = \frac{5 \text{ inches}}{n \text{ miles}}$

5 inches on the map represent 375 miles.

5. $\frac{2 \text{ inches}}{150 \text{ miles}} = \frac{n \text{ inches}}{375 \text{ miles}}$

375 miles are represented by 5 inches.

6. $\frac{1.5 \text{ inches}}{20 \text{ miles}} = \frac{6 \text{ inches}}{n \text{ miles}}$

6 inches on the map represent 80 miles.

7. $\frac{1 \text{ inch}}{150 \text{ miles}} = \frac{3.5 \text{ inches}}{n \text{ miles}}$

It is 525 miles between two cities that are 3.5 inches apart.

8. $\frac{.5 \text{ inch}}{50 \text{ miles}} = \frac{7 \text{ inches}}{n \text{ miles}}$

7 inches on the scale equal 700 miles.

Page 70: Make a Chart

1.

hours of sleep	9	n
days	1	365

2.

dollars earned	10	155
dollars saved	1	n

3.

inches of snow	1.5	n
hours	1	4

4.

swimmers	6	n
number of kids	10	350

Page 71: Turn Charts into Proportions

1.

dollar bills	1	n
dimes	10	130

$\frac{1}{10} = \frac{n}{130}$

$n = 13$ dollar bills

2.

tapes bought	6	9
free tapes	2	n

$\frac{6}{2} = \frac{9}{n}$

$n = 3$ free tapes

3.

boxes of cereal	1.5	n
bags of pretzels	1	3

$\frac{1.5}{1} = \frac{n}{3}$

$n = 4.5$ boxes of cereal needed

4.

aspirin tablets	2	n
hours	4	24

$\frac{2}{4} = \frac{n}{24}$

$n = 12$ aspirin tablets in 24 hours

Page 72: Proportion in Measurement

1.

pound	1	n
ounces	16	128

$\frac{1 \text{ pound}}{16 \text{ ounces}} = \frac{n \text{ pounds}}{128 \text{ ounces}}$

$n = 8$ pounds

2.

pecks	4	n
bushels	1	52

$\frac{4 \text{ pecks}}{1 \text{ bushel}} = \frac{n \text{ pecks}}{52 \text{ bushels}}$

$n = 208$ pecks

3.

grams	1	564
ounces	.04	n

$\frac{1 \text{ gram}}{.04 \text{ ounce}} = \frac{564 \text{ grams}}{n \text{ ounces}}$

$n = 22.56$ ounces

4.

centimeters	100	n
meters	1	655

$\frac{100 \text{ centimeters}}{1 \text{ meter}} = \frac{n \text{ centimeters}}{655 \text{ meters}}$

$n = 65{,}500$ centimeters

Page 73: Proportions in Sports

1.

miles	120	n
days	3	7

$\frac{120 \text{ miles}}{3 \text{ days}} = \frac{n \text{ miles}}{7 \text{ days}}$

$n = 280$ miles

2.

mile	8.4	n
hours	2	3

$\frac{8.4 \text{ miles}}{2 \text{ hours}} = \frac{n \text{ miles}}{3 \text{ hours}}$

$n = 12.6$ miles

3.

home runs	4	n
games	6	9

$\frac{4 \text{ home runs}}{6 \text{ games}} = \frac{n \text{ home runs}}{9 \text{ games}}$

$n = 6$ home runs

4.

points	92	n
games	4	18

$\frac{92 \text{ points}}{4 \text{ games}} = \frac{n \text{ points}}{18 \text{ games}}$

$n = 414$ points

Page 74: Proportion Problem-Solving Review

1. 8
2. 9
3. $1.75
4. 412
5. 12
6. 35
7. 2,500
8. 240

Page 75: Cumulative Review

1. 18:36 = 1:2
2. 20:60 = 1:3
3. 8:3
4. 54 blue cars
5. $.23
6. 72 minutes

Page 76: Posttest

1. $\frac{19}{57} = \frac{1}{3}$
2. $m = 35$
3. 120 miles
4. $x = 1\frac{1}{3}$
5. 5:6
6. 9:16
7. 5:9
8. 3:22
9. $a = 1.2$
10. 7 weeks
11. 8 minutes
12. 1:8
13. $c = 10$
14. $8.70
15. 3:20
16. $m = 4\frac{4}{5}$ or 4.8
17. 5:4
18. 17:40
19. $3\frac{1}{2}$ hours or 3.5 hours
20. 1:4 or 1 to 4

Posttest Evaluation Chart

If a student misses one or more problems in a skill area, assign a review of the practice pages for that skill.

Skill Area	Posttest Problem #	Skill Section	Review Page
Meaning of Ratio	1, 5, 8	7–16	17
Ratio Applications	7, 12, 17	18–32	33
Ratio Problem Solving	15, 18, 20	34–41	42
Meaning of Proportion	2, 4, 9, 13, 16	43–55	56
Proportion Applications	All	57–63	64
Proportion Applications & Problem Solving	3, 6, 10, 11, 14, 19	65–73	74

THE MEANING OF PERCENT

It is important to stress that percent is a special ratio involving parts out of 100. *Per* means *for every* and *cent* means *100*. When writing a ratio comparing a number to 100, the symbol % is used to represent the ratio. For example, 25% means 25 out of every 100.

Emphasis should be placed on the basic concepts of percents using patterns, not just mechanical manipulation of the rules.

Student Glossary

Acquainting students with definitions of key math terms and life-skills concepts will enhance their mastery of the materials. Below are words defined in the student text. A glossary is provided at the end of the student text and on page 243 of the *Teacher's Resource Guide and Answer Key.*

attendance
budget
commission
denominator
discount
earnings
equivalent
improper fraction
mixed number
number relation symbol
percent
ratio
remainder
rename
simplify (reduce)
tip

STUDENT PAGE
10
Meaning of Percent

Shade Your Name

Provide graph paper for students to mark off 10 × 10 squares. Have students shade their initials using the grids similar to the example below. Discuss with your students questions similar to the following.

Sample questions for each shaded box:

What ratio is represented by the shaded regions? ______ : ______

Write this ratio as a percent. ______%

What percent is not shaded? ______%

What percent shows the sum of the shaded and the nonshaded parts? ______%

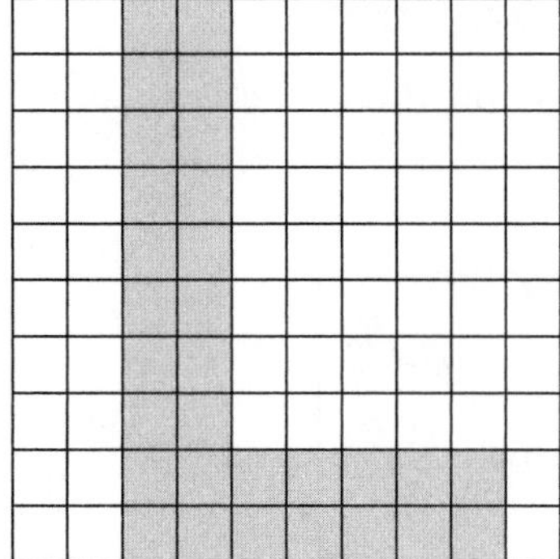

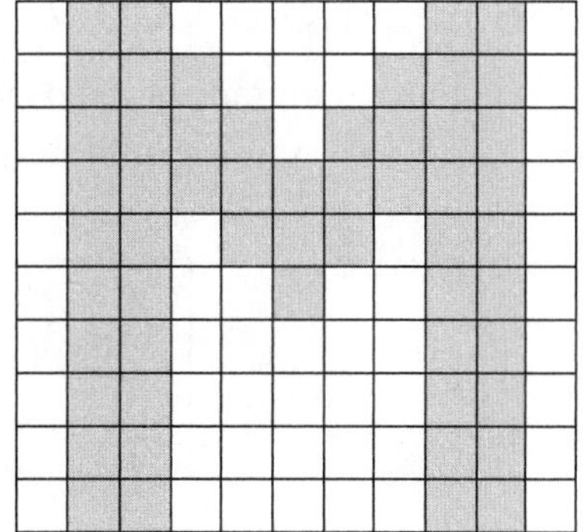

STUDENT PAGE
12
Meaning of Percent

Using Percents with Measures

Ask students to draw a 10-inch line segment. Let this line represent 100%. Have students divide the line segment into 10 equal 1-inch divisions.

Ask them to draw these lines:

50% of a 10-inch line (Answer: 5 inches)

20% of a 10-inch line (Answer: 2 inches)

70% of a 10-inch line (Answer: 7 inches)

80% of a 10-inch line (Answer: 8 inches)

75% of a 10-inch line (Answer: $7\frac{1}{2}$ inches)

Have them continue this with many more lines of different lengths.

STUDENT PAGE
19
Meaning of Percent

Fold and Label Paper Strips

Ask students to cut out paper strips, fold them, and label percents.

Emphasize that the whole strip is: $100\% = \frac{100}{100} = 1 =$ the whole thing.

$\frac{1}{2} = 50\%$	

$\frac{1}{4} = 25\%$			

$\frac{1}{3} = 30\%$		

STUDENT PAGE
33
Fractions and Percents

Unscramble Equivalent Values

Place many commonly used percents and their fractional equivalents in a box and have your students unscramble and pair them.

EXAMPLE

$\frac{1}{4}$ 10% $\frac{3}{4}$ 25%
$\frac{1}{3}$ 50% $\frac{1}{5}$
$\frac{1}{10}$ 75% $\frac{1}{2}$ 20%

STUDENT PAGE
36
Percents Greater than 100

Percents Greater than 100%

The concept of a percent being greater than 100 can be very confusing for students. You can relate this idea to concrete situations to help students visualize the concepts.

SAMPLE QUESTIONS

Can you have 100% attendance in class? What does this mean? Can you have 200% attendance in class?

What does a price increase of 100% mean?

Can a price increase 400%?

Is it possible to have a 100% discount?

Discuss real life situations where percents are greater than 100. How many situations can the students list?

STUDENT PAGE
37
Percents Greater than 100

More Percents Greater than 100%

To help your students understand percents greater than 100, give them problems similar to the examples below.

EXAMPLES

If □ = 100%, then □□ = 200%

If □□ = 100%, then □□□ = 150%

If □□□□ = 100%, then □□□□□□□ = _____%

If XXXXX = 100%, then XXXXXXXXXX = _____%

If 3 = 100%, then 6 = _____%

If 6 = 100%, then 9 = _____%

STUDENT PAGE
53
Decimals, Fractions, and Percents

Relating Money to Percent

Have students complete a chart similar to the one below.

Fraction of a Dollar	Amount Shown as a Decimal	Ratio to 100 in Fraction Form	Percent
$\frac{1}{4}$	\$. __ __	$\frac{\square}{100}$	___ %
	\$. 5 0		
		$\frac{5}{100}$	
	\$. 1 0		
$\frac{1}{5}$			

STUDENT PAGE
55
Decimals, Fractions, and Percents

Compare Mentally

On the board or overhead projector, write problems that are similar to the following and have students say which of the two numbers is smaller. Have students support their position and not just guess.

A 20% or $\frac{1}{10}$

B $\frac{1}{5}$ or 5%

C $\frac{3}{4}$ or 80%

D 75% or $\frac{4}{5}$

E .8 or 81%

F 1 or 1%

G 5 or 5%

H 30% or $\frac{1}{4}$

I $\frac{1}{3}$ or 30%

J $\frac{1}{10}$ or 20%

STUDENT PAGE
68
Percent Problem Solving

Newspaper Percents

Have students bring in newspapers, supermarket fliers, or other advertisement material that shows percents. In groups, students should write and solve percent problems that relate to their daily lives, by practicing changing percents to fractions and vice versa.

EXAMPLE

25% OFF THIS WEEKEND ONLY!!!!

What is 25% written as a fraction? $\frac{25}{100}$ or $\frac{1}{4}$

If a shirt regularly sold for \$20, what is 25% of 20? \$5

STUDENT PAGE
69
Percent Problem Solving

Using Circles to Develop Meaning

Provide students with a circle marked off in 100 equal divisions.

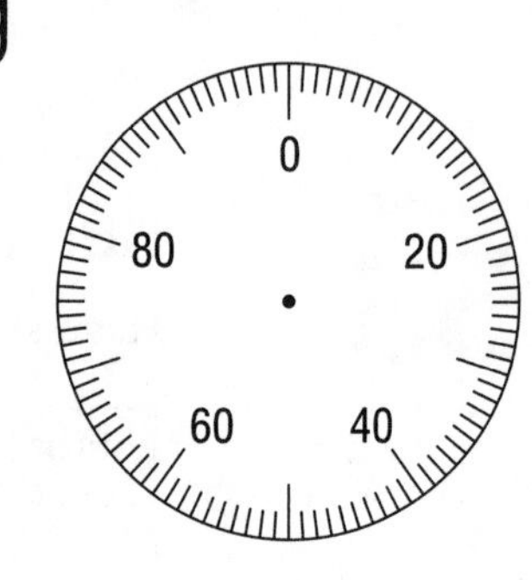

Have students use the circle to find the percent of regions of other circles.

EXAMPLE

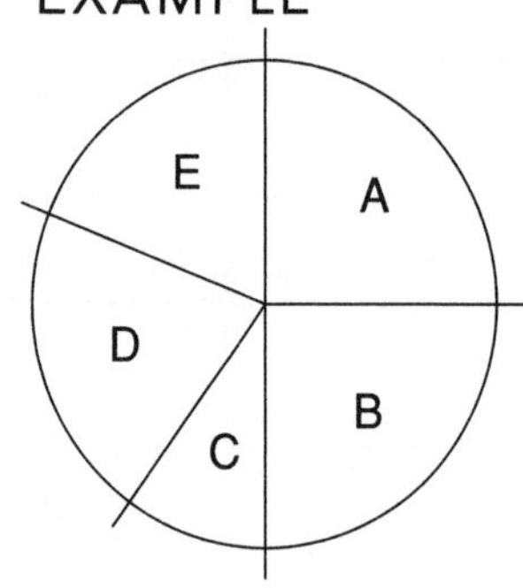

SAMPLE QUESTIONS

1 What percent of the entire circle does each region represent?

A ______ **D** ______

B ______ **E** ______

C ______

2 What is the sum of the percents represented by A, B, C, D, and E? ______

Page 4: Pretest

1. 90
2. 75%
3. 100
4. 75%
5. $\frac{3}{10}$
6. $\frac{44}{100} = \frac{11}{25}$
7. 35%
8. $37\frac{1}{2}\%$
9. $2\frac{2}{5}$
10. 325%
11. 500%
12. 5%
13. .25
14. 80%
15. .68
16. 275%
17. 25%
18. 70%
19. 10%
20. 75%

Pretest Evaluation Chart

If a student misses any problems in a skill area, assign the practice pages for that skill. However, students may need to complete all practice pages to reinforce areas of weakness.

Skill Area	Pretest Problem #	Skill Section	Review Page
Meaning of Percent	1, 2, 3	7–17	18
Fractions and Percent	4, 5, 6, 7, 8	19–34 61–65	35 66
Percents Greater than 100	9, 10, 11	36–40	41
Percents and Decimals	13, 15, 16	42–50	51
Decimals, Fractions, and Percents	12, 14	52–58	59
Percent Problem Solving	17, 18, 19, 20	67–73	74

Page 7: What Is Percent?

1. 45
2. $\frac{45}{100}$
3. 45%
4. 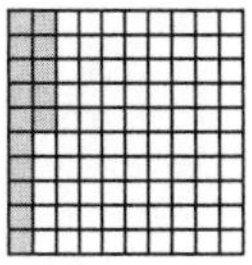
5. $\frac{15}{100}$
6. 15%
7. 25
8. 15
9. 75
10. 90
11. 5
12. 1

Page 8: Shade the Percents

1. 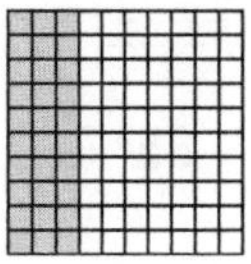
2. $\frac{30}{100}$
3. 30%
4. 70%
5. 100%
6. 
7. $\frac{65}{100}$
8. 65%
9. 35%
10. 100%
11. 
12. $\frac{87}{100}$
13. 87%
14. 13%
15. 100%

Page 9: Understanding Percent

1. 45 out of 100 or $\frac{45}{100}$
2. 60 out of 100 or $\frac{60}{100}$
3. 55 out of 100 or $\frac{55}{100}$
4. 5 out of 100 or $\frac{5}{100}$
5. $\frac{95}{100}$ = 95%
6. $\frac{100}{100}$ = 100%
7. $\frac{50}{100}$ = 50%
8. $\frac{75}{100}$ = 75%
9. 33 out of 100 or 33%
10. 20 out of 100 or 20%
11. 10 out of 100 or 10%
12. 18 out of 100 or 18%

Page 10: Percent Means Parts Out of 100

1. $\frac{24}{100} = 24\%$

2. $\frac{52}{100} = 52\%$

3. $\frac{48}{100} = 48\%$

4. $\frac{69}{100} = 69\%$

5. $\frac{33}{100} = 33\%$

6. $\frac{60}{100} = 60\%$

Page 11: Shade the Squares

1. 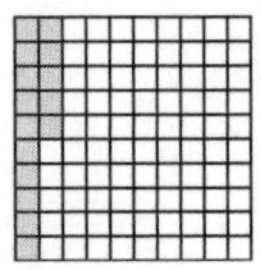

$\frac{15}{100} = 15\%$

2. 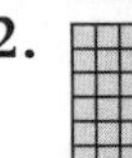 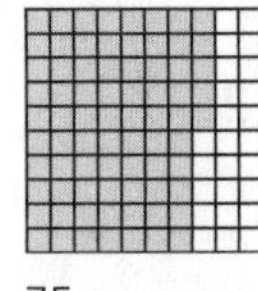

$\frac{75}{100} = 75\%$

3. 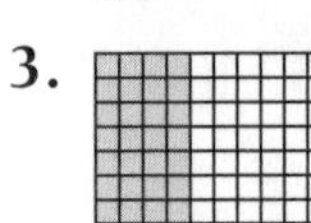

$\frac{37}{100} = 37\%$

4. 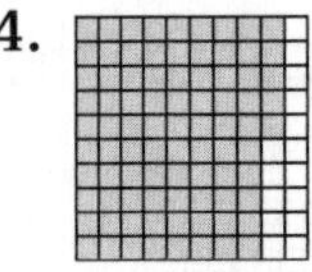

$\frac{85}{100} = 85\%$

5. 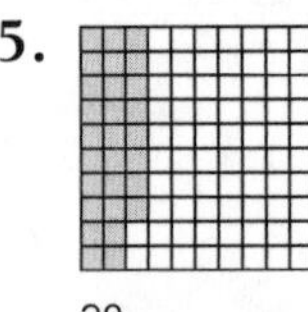

$\frac{28}{100} = 28\%$

6. 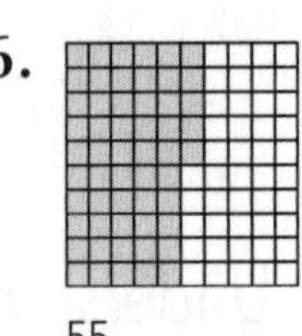

$\frac{55}{100} = 55\%$

Page 12: Number Lines

1. 70% — 0% 50% 100% — 70%

2. 15% — 0% 50% 100% — 15%

3. 95% — 0% 50% 100% — 95%

4. 35% — 0% 50% 100% — 35%

5. 20% — 0% 50% 100% — 20%

Page 13: One Hundred Percent Equals One

A. 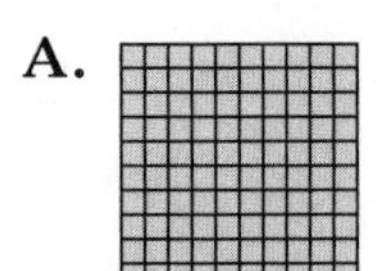

B. 100%

C. 100%

1. 50%
2. 25%
3. 91%
4. 55%
5. 36%
6. 45%
7. 30%
8. 40%
9. 55%

Page 14: Fifty Percent

A. $\frac{50}{100} = 50\%$

1. a)
b) 3
c) 3

2. a) 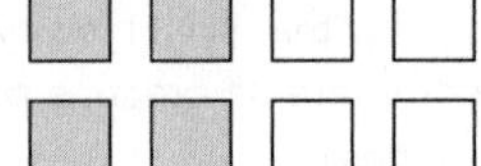
b) 6
c) 6

3. a) 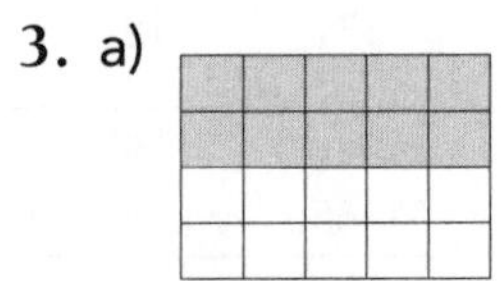
b) 10
c) 10

4. 5
5. 8
6. 15
7. 12
8. 9
9. 25
10. 20

THE MEANING OF PERCENT

Page 15: Twenty-Five Percent

A. $\frac{25}{100} = 25\%$

1. a) 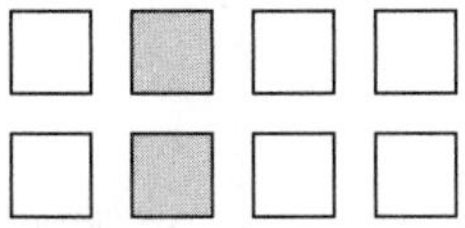

b) 2

c) 2

2. a) 

b) 3

c) 3

3. 8
4. 4
5. 5
6. 10
7. 6
8. 9
9. 1
10. 7

Page 16: What Does the Percent Mean?

1. 1 out of 2
2. 5 out of 10
3. 10 out of 20
4. 15 out of 30
5. 9 out of 18
6. 25 out of 50
7. 1 out of 4
8. 2 out of 8
9. 5 out of 20
10. 10 out of 40
11. 3 out of 12
12. 4 out of 16
13. 1 out of 10
14. 3 out of 30
15. 2 out of 20
16. 8 out of 80
17. 5 out of 50
18. 9 out of 90
19. 1 out of 5
20. 5 out of 25
21. 2 out of 10
22. 7 out of 35
23. 3 out of 15
24. 9 out of 45

Page 17: More Percent Meanings

1. 3 out of 5
2. 6 out of 10
3. 12 out of 20
4. 24 out of 40
5. 30 out of 50
6. 36 out of 60
7. 2 out of 5
8. 20 out of 50
9. 4 out of 10
10. 16 out of 40
11. 8 out of 20
12. 12 out of 30
13. 4 out of 5
14. 8 out of 10
15. 12 out of 15
16. 20 out of 25
17. 36 out of 45
18. 28 out of 35
19. 3 out of 4
20. 6 out of 8
21. 15 out of 20
22. 21 out of 28
23. 9 out of 12
24. 18 out of 24

Page 18: Meaning of Percent Review

1. 35
2. 25
3. 45%
4. 52%
5. 

6. 13
7. 37
8. 9
9. 11
10. 3

Page 19: Common Fractions and Percents

1. $33\frac{1}{3}\%$
2. 50%
3. 25%
4. 75%
5. 10%
6. $66\frac{2}{3}\%$
7. 25%
8. 50%
9. $33\frac{1}{3}\%$

Page 20: Rename as a Percent

1. $\frac{1}{2} = \frac{50}{100} = 50\%$
2. $\frac{1}{4} = \frac{25}{100} = 25\%$
3. $\frac{1}{5} = \frac{20}{100} = 20\%$
4. $\frac{1}{10} = \frac{10}{100} = 10\%$

Page 21: Ratios to Percents

1. 3
2. $\frac{3}{10}$
3. 30; $\frac{30}{100} = 30\%$
4. 7
5. 70%
6. 5
7. $\frac{5}{20}$
8. 25; $\frac{25}{100} = 25\%$
9. 15
10. 75%

Page 22: Apply Your Skills

1. $\frac{1 \times 50}{2 \times 50} = \frac{50}{100} = 50\%$
2. $\frac{1 \times 25}{4 \times 25} = \frac{25}{100} = 25\%$
3. $\frac{4 \times 10}{10 \times 10} = \frac{40}{100} = 40\%$
4. $\frac{3 \times 20}{5 \times 20} = \frac{60}{100} = 60\%$

Page 23: Use the Pictures

1. $\frac{1}{4} = \frac{25}{100} = 25\%$
2. $\frac{3}{4} = \frac{75}{100} = 75\%$
3. $\frac{4}{5} = \frac{80}{100} = 80\%$
4. $\frac{3}{5} = \frac{60}{100} = 60\%$
5. $\frac{6}{8} = \frac{3}{4} = \frac{75}{100} = 75\%$
6. $\frac{2}{8} = \frac{1}{4} = \frac{25}{100} = 25\%$
7. $\frac{3}{6} = \frac{1}{2} = \frac{50}{100} = 50\%$
8. $\frac{4}{16} = \frac{1}{4} = \frac{25}{100} = 25\%$

Page 24: Shade the Percents

1. 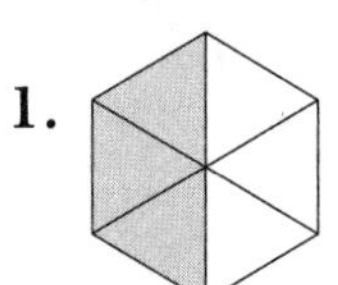

$\frac{1}{2}$ or 50%

2. 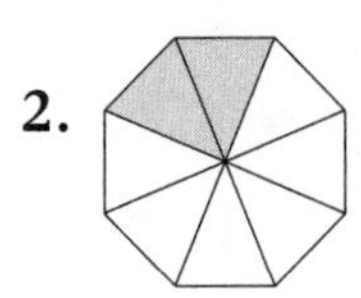

$\frac{1}{4}$ or 25%

3. 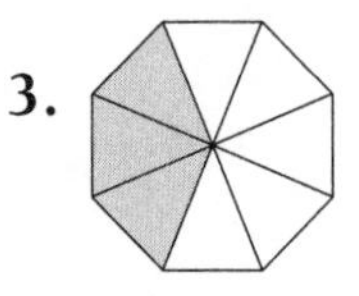

$\frac{3}{8}$ or $37\frac{1}{2}\%$

4. 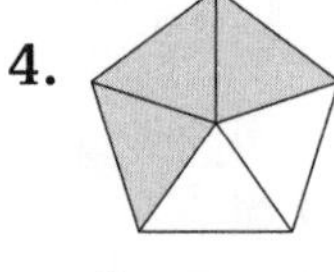

$\frac{3}{5}$ or 60%

5. 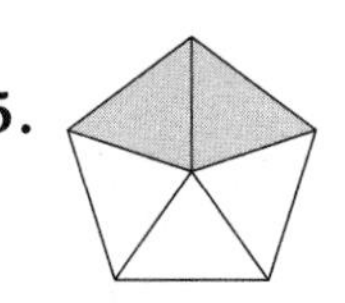

$\frac{2}{5}$ or 40%

6. 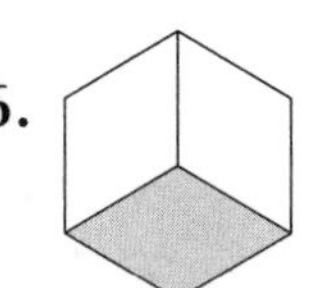

$\frac{1}{3}$ or $33\frac{1}{3}\%$

7. 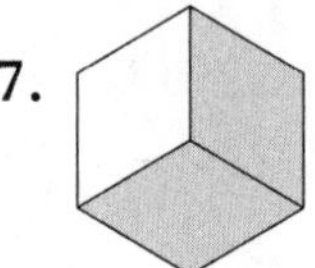

$\frac{2}{3}$ or $66\frac{2}{3}\%$

8. 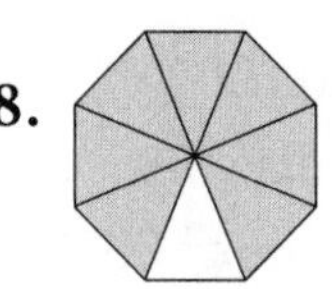

$\frac{7}{8}$ or $87\frac{1}{2}\%$

Page 25: Name the Percent

1. $\frac{7}{10} = \frac{70}{100} = 70\%$
2. $\frac{4}{20} = \frac{20}{100} = 20\%$
3. $\frac{4}{5} = \frac{80}{100} = 80\%$
4. $\frac{10}{25} = \frac{40}{100} = 40\%$

Page 26: Think It Through

1\.

2\.

3\. $\frac{15}{20} = \frac{75}{100} = 75\%$

4\. $\frac{4}{5} = \frac{80}{100} = 80\%$

5\. $\frac{2}{8} = \frac{1}{4} = \frac{25}{100} = 25\%$

Page 27: Change Fractions to Percents

A. 50%

B. 40%

C. 75% $\frac{3}{4} \times \frac{25}{25} = \frac{75}{100} = 75\%$

1\. $\frac{2}{10} = \frac{20}{100} = 20\%$

2\. $\frac{3}{5} = \frac{60}{100} = 60\%$

3\. $\frac{1}{4} = \frac{25}{100} = 25\%$

4\. $\frac{7}{10} = \frac{70}{100} = 70\%$

5\. $\frac{1}{25} = \frac{4}{100} = 4\%$

6\. $\frac{4}{5} = \frac{80}{100} = 80\%$

7\. $\frac{1}{20} = \frac{5}{100} = 5\%$

8\. $\frac{5}{5} = \frac{100}{100} = 100\%$

9\. $\frac{7}{20} = \frac{35}{100} = 35\%$

10\. $\frac{9}{50} = \frac{18}{100} = 18\%$

Page 28: Percent Wise

1\. 80%
2\. 20%
3\. 30%
4\. 70%
5\. 65%
6\. 35%
7\. 75%
8\. 25%

Page 29: Show What Percent Is Shaded

1\. a) 50%
b) 50%

2\. a) 75%
b) 25%

3\. a) 60%
b) 40%

4\. a) 60%
b) 40%

5\. a) 100%
b) 0%

6\. a) 36%
b) 64%

Page 30: Rename Percents as Fractions

1\. $75\% = \frac{75}{100} = \frac{3}{4}$

2\. $80\% = \frac{80}{100} = \frac{4}{5}$

3\. $60\% = \frac{60}{100} = \frac{3}{5}$

4\. $10\% = \frac{10}{100} = \frac{1}{10}$

Page 31: Percents Are Special Ratios

1\. a) 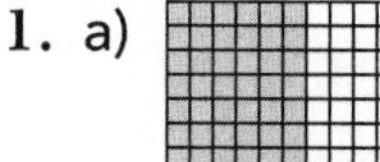

b) 60%

c) $\frac{60}{100}$

d) $\frac{60}{100} = \frac{3}{5}$

2\. $\frac{7}{20}$

3\. $\frac{2}{5}$

4\. $\frac{9}{20}$

Page 32: Change Percents to Fractions

1\. $\frac{1}{5}$

2\. $\frac{1}{10}$

3\. $\frac{30}{100} = \frac{3}{10}$

4\. $\frac{50}{100} = \frac{1}{2}$

5\. $\frac{70}{100} = \frac{7}{10}$

6\. $\frac{4}{100} = \frac{1}{25}$

7\. $\frac{90}{100} = \frac{9}{10}$

8\. $\frac{3}{4}$

9\. $\frac{1}{20}$

10\. $\frac{45}{100} = \frac{9}{20}$

11\. $\frac{80}{100} = \frac{4}{5}$

12\. $\frac{15}{100} = \frac{3}{20}$

13\. $\frac{60}{100} = \frac{3}{5}$

14\. $\frac{40}{100} = \frac{2}{5}$

Page 33: Practice Changing Percents to Fractions

1. $\frac{25}{100} = \frac{1}{4}$
2. $\frac{8}{100} = \frac{2}{25}$
3. $\frac{15}{100} = \frac{3}{20}$
4. $\frac{4}{100} = \frac{1}{25}$
5. $\frac{75}{100} = \frac{3}{4}$
6. $\frac{20}{100} = \frac{1}{5}$
7. $\frac{90}{100} = \frac{9}{10}$
8. $\frac{6}{100} = \frac{3}{50}$
9. $\frac{12}{100} = \frac{3}{25}$
10. $\frac{10}{100} = \frac{1}{10}$
11. $\frac{45}{100} = \frac{9}{20}$
12. $\frac{1}{100} = \frac{1}{100}$
13. $\frac{17}{100} = \frac{17}{100}$
14. $\frac{22}{100} = \frac{11}{50}$
15. $\frac{60}{100} = \frac{3}{5}$
16. $\frac{36}{100} = \frac{9}{25}$
17. $\frac{9}{100} = \frac{9}{100}$
18. $\frac{28}{100} = \frac{7}{25}$

Page 34: Common Equivalents

1. $\frac{1}{4}$
2. $\frac{1}{5}$
3. $\frac{1}{2}$
4. $\frac{1}{10}$
5. $\frac{2}{3}$
6. $\frac{1}{8}$
7. $\frac{3}{4}$
8. $\frac{1}{3}$
9. $66\frac{2}{3}\%$
10. $12\frac{1}{2}\%$
11. 75%
12. 20%
13. 10%
14. 25%
15. $33\frac{1}{3}\%$
16. 50%

Page 35: Fractions and Percents Review

1. 30%
2. $\frac{6}{16} = 37.5\%$
3. 70%
4. $\frac{3}{5} = \frac{60}{100} = 60\%$
5. $\frac{17}{25} = \frac{68}{100} = 68\%$
6. a) $\frac{60}{100} = \frac{3}{5}$
   b) $\frac{45}{100} = \frac{9}{20}$
7. a) $\frac{22}{100} = \frac{11}{50}$
   b) $\frac{38}{100} = \frac{19}{50}$
8. a) $\frac{9}{100}$
   b) $\frac{12}{100} = \frac{6}{50} = \frac{3}{25}$
9. a) $\frac{1}{4}$
   b) $\frac{1}{3}$
10. a) 12.5%
    b) 20%

Page 36: Numbers Greater Than 100%

1.

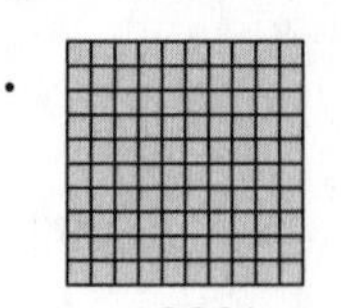 

100% + 50% = 150%

$1 + \frac{1}{2} = 1\frac{1}{2}$

2.

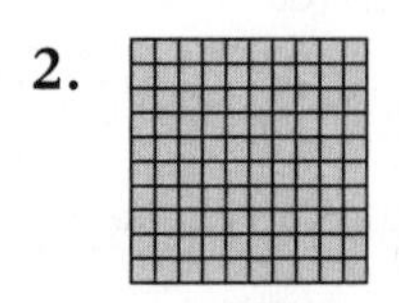 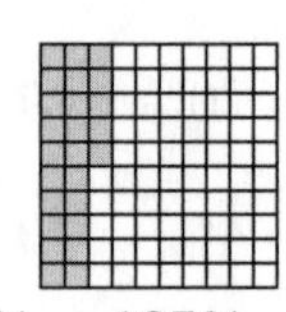

100% + 25% = 125%

$1 + \frac{1}{4} = 1\frac{1}{4}$

3.

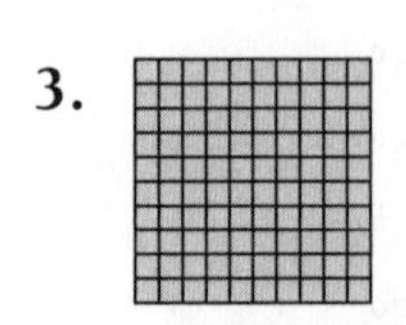 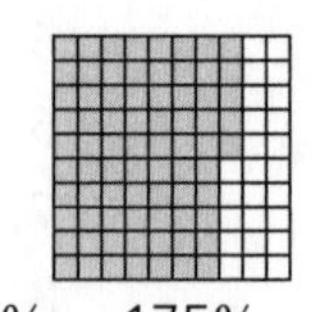

100% + 75% = 175%

$1 + \frac{3}{4} = 1\frac{3}{4}$

THE MEANING OF PERCENT

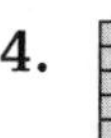

4. 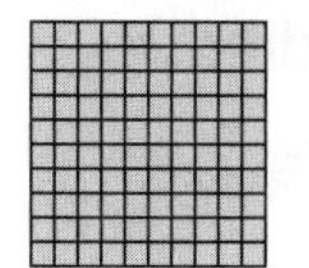 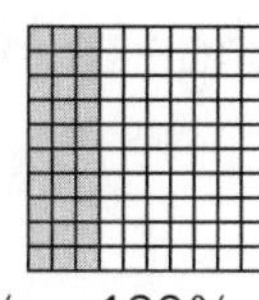

100% + 30% = 130%

$1 + \frac{3}{10} = 1\frac{3}{10}$

5. 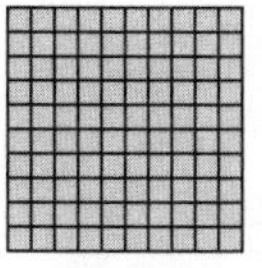 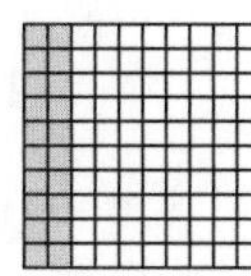

100% + 20% = 120%

$1 + \frac{1}{5} = 1\frac{1}{5}$

6. 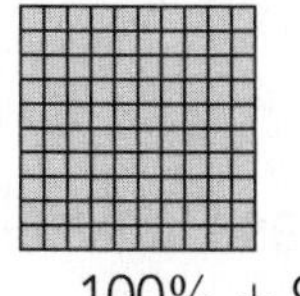 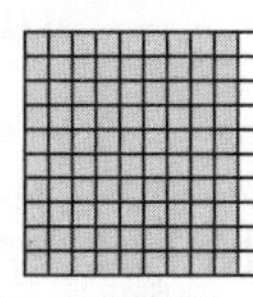

100% + 90% = 190%

$1 + \frac{9}{10} = 1\frac{9}{10}$

9. 225% = 200% + 25%

$= 2 + \frac{25}{100}$

$= 2\frac{1}{4}$

10. 460% = 400% + 60%

$= 4 + \frac{60}{100}$

$= 4\frac{3}{5}$

11. 270% = 200% + 70%

$= 2 + \frac{70}{100}$

$= 2\frac{7}{10}$

12. 580% = 500% + 80%

$= 5 + \frac{80}{100}$

$= 5\frac{4}{5}$

Page 37: Change Percents to Mixed Numbers

1. $1 = \frac{100}{100} = 100\%$
2. $2 = \frac{200}{100} = 200\%$
3. $5 = \frac{500}{100} = 500\%$
4. $3 = \frac{300}{100} = 300\%$
5. $4 = \frac{400}{100} = 400\%$
6. $8 = \frac{800}{100} = 800\%$
7. 310% = 300% + 10%

   $= 3 + \frac{10}{100}$

   $= 3 + \frac{1}{10} = 3\frac{1}{10}$
8. 450% = 400% + 50%

   $= 4 + \frac{50}{100}$

   $= 4 + \frac{1}{2} = 4\frac{1}{2}$

Page 38: Practice Helps

1. $6 = \frac{600}{100} = 600\%$
2. $9 = \frac{900}{100} = 900\%$
3. $7 = \frac{700}{100} = 700\%$
4. $2 = \frac{200}{100} = 200\%$
5. $4 = \frac{400}{100} = 400\%$
6. $460\% = 4\frac{3}{5}$
7. $380\% = 3\frac{4}{5}$
8. $250\% = 2\frac{1}{2}$
9. $125\% = 1\frac{1}{4}$
10. $880\% = 8\frac{4}{5}$

Page 39: Mixed Practice

1. $\frac{20}{100} = \frac{1}{5}$
2. $2 + \frac{75}{100} = 2\frac{3}{4}$
3. $3 + \frac{50}{100} = 3\frac{1}{2}$
4. $\frac{45}{100} = \frac{9}{20}$
5. $\frac{5}{100} = \frac{1}{20}$
6. $1 + \frac{60}{100} = 1\frac{3}{5}$
7. $\frac{8}{100} = \frac{2}{25}$
8. $6\frac{1}{4}$
9. $\frac{1}{10}$
10. $4\frac{1}{5}$
11. $\frac{4}{5}$
12. $\frac{1}{25}$
13. $1\frac{1}{20}$
14. $\frac{17}{20}$

Page 40: More Mixed Practice

1. 15%
2. 25%
3. 70%
4. 25%
5. $2\frac{3}{4}$
6. 10%
7. $\frac{3}{10}$
8. 70%
9. 750%
10. 50%
11. 20%
12. 40%
13. 50%
14. $\frac{9}{10}$
15. 25%

Page 41: Percent Review

1.
2. 50%
3. 25%
4. 75%
5. 20%
6. 610%
7. 9
8. 8
9. $\frac{20}{100} = \frac{1}{5}$
10. $\frac{50}{100} = \frac{1}{2}$
11. $2\frac{50}{100} = 2\frac{1}{2}$
12. $\frac{10}{100} = \frac{1}{10}$
13. 80%
14. 20%
15. 50%
16. 100%
17. 10%
18. 25%

Page 42: Change Percents to Decimals

1. 10 parts out of 100 = $\frac{10}{100}$ = .10
2. 85 parts out of 100 = $\frac{85}{100}$ = .85
3. 33 parts out of 100 = $\frac{33}{100}$ = .33
4. 40 parts out of 100 = $\frac{40}{100}$ = .40
5. 15 parts out of 100 = $\frac{15}{100}$ = .15
6. 5 parts out of 100 = $\frac{5}{100}$ = .05
7. 3 parts out of 100 = $\frac{3}{100}$ = .03
8. 1 part out of 100 = $\frac{1}{100}$ = .01

Page 43: Change to Hundredths

1. $\frac{15}{100}$ = .15
2. $\frac{18}{100}$ = .18
3. $\frac{39}{100}$ = .39
4. $\frac{7}{100}$ = .07
5. $\frac{20}{100}$ = .20
6. $\frac{9}{100}$ = .09
7. $\frac{75}{100}$ = .75
8. $\frac{65}{100}$ = .65
9. $\frac{5}{100}$ = .05
10. $\frac{98}{100}$ = .98
11. $\frac{67}{100}$ = .67
12. $\frac{12}{100}$ = .12
13. $\frac{10}{100}$ = .10
14. $\frac{4}{100}$ = .04

Page 44: Write Percents as Decimals

1. .75
2. .36
3. .16
4. .44
5. .70
6. .93
7. .08
8. .82
9. .10
10. .05
11. .07
12. .22

Page 45: Change Percents to Mixed Decimals

1. 3.25
2. 4.01
3. 1.36
4. 2.19
5. 1.50
6. 6.00
7. 3.00
8. 2.75
9. 1.00
10. 1.92
11. 5.00
12. 4.50

Page 46: Mixed Practice

1. .054
2. .217
3. .0368
4. .45
5. 3.00
6. .095
7. 3.48
8. .026
9. 3.50
10. .016
11. .543
12. .19

Page 47: Change Decimals to Percents

1. $\frac{65}{100}$ = 65 hundredths = 65%
2. $\frac{19}{100}$ = 19 hundredths = 19%
3. $\frac{33}{100}$ = 33 hundredths = 33%
4. $\frac{80}{100}$ = 80 hundredths = 80%
5. $\frac{1}{100}$ = 1 hundredth = 1%
6. $\frac{68}{100}$ = 68 hundredths = 68%
7. $\frac{75}{100}$ = 75 hundredths = 75%
8. $\frac{66}{100}$ = 66 hundredths = 66%
9. $\frac{8}{100}$ = 8 hundredths = 8%
10. $\frac{86}{100}$ = 86 hundredths = 86%

Page 48: Change One-Place Decimals to Percents

1. $\frac{8}{10} \times \frac{10}{10} = \frac{80}{100}$ = 80%
2. $\frac{1}{10} \times \frac{10}{10} = \frac{10}{100}$ = 10%
3. $\frac{6}{10} \times \frac{10}{10} = \frac{60}{100}$ = 60%
4. $\frac{5}{10} \times \frac{10}{10} = \frac{50}{100}$ = 50%
5. $\frac{3}{10} \times \frac{10}{10} = \frac{30}{100}$ = 30%
6. $\frac{2}{10} \times \frac{10}{10} = \frac{20}{100}$ = 20%
7. $\frac{7}{10} \times \frac{10}{10} = \frac{70}{100}$ = 70%
8. $\frac{9}{10} \times \frac{10}{10} = \frac{90}{100}$ = 90%

Page 49: Relating Decimals to Percents

1. 26%
2. 35.4%
3. 148%
4. 88.7%
5. 36%
6. 10.9%
7. 44%
8. 15.6%
9. 78%
10. 292%
11. 57%
12. 113%

Page 50: Add Zeros When Necessary

1. 30%
2. 50%
3. 340%
4. 910%
5. 70%
6. 800%
7. 80%
8. 600%
9. 40%
10. 590%
11. 10%
12. 320%

Page 51: Percents and Decimals Review

1. 1.57
2. 2.28
3. 9.11
4. 8.13
5. 4.56
6. 30%
7. 50%
8. 260%
9. 540%
10. 800%

Page 52: Put It All Together

1. a) $\frac{1}{2}$
   b) .50
   c) 50%
2. a) $\frac{3}{4}$
   b) .75
   c) 75%
3. a) $\frac{1}{5}$
   b) .20
   c) 20%
4. a) $\frac{7}{10}$
   b) .70
   c) 70%
5. a) $\frac{4}{5}$
   b) .80
   c) 80%
6. a) $\frac{3}{5}$
   b) .60
   c) 60%

Page 53: Equivalents

	Fraction (Simplified)	Fraction (Hundredths)	Decimal	Percent
1.	$\frac{1}{2}$	$\frac{50}{100}$	.50	50%
2.	$\frac{1}{4}$	$\frac{25}{100}$	.25	25%
3.	$\frac{3}{4}$	$\frac{75}{100}$	.75	75%
4.	$\frac{1}{10}$	$\frac{10}{100}$	.10	10%
5.	$\frac{7}{10}$	$\frac{70}{100}$	.70	70%
6.	$\frac{3}{5}$	$\frac{60}{100}$	.60	60%
7.	$\frac{2}{5}$	$\frac{40}{100}$	.40	40%
8.	$\frac{3}{10}$	$\frac{30}{100}$	.30	30%
9.	$\frac{1}{50}$	$\frac{2}{100}$	.02	2%
10.	$\frac{1}{5}$	$\frac{20}{100}$	.20	20%

Page 54: Comparing Percents, Fractions, and Decimals

1. =
2. <
3. >
4. <
5. >
6. <
7. <
8. >
9. =
10. >
11. <
12. =

Page 55: Using Proportions

Example 2

Step 3: 200

Step 4: 200

$66\frac{2}{3} = n$

B. $66\frac{2}{3}\%$

Page 56: Use Cross Products

1. $33\frac{1}{3}\%$
2. $16\frac{2}{3}\%$
3. $87\frac{1}{2}\%$
4. $62\frac{1}{2}\%$
5. $11\frac{1}{9}\%$
6. $83\frac{1}{3}\%$

Page 57: Find the Percent

1. $12\frac{1}{2} = n$
   $12\frac{1}{2}\% = \frac{1}{8}$
2. $37\frac{1}{2}\% = \frac{3}{8}$
3. $66\frac{2}{3}\% = \frac{2}{3}$
4. $33\frac{1}{3}\% = \frac{1}{3}$

Page 58: Picture These Percents

1. $\frac{7}{10} = \frac{70}{100} = 70\%$
2. $\frac{3}{8} = \frac{37\frac{1}{2}}{100} = 37\frac{1}{2}\%$
3. $\frac{10}{20} = \frac{50}{100} = 50\%$
4. $\frac{3}{4} = \frac{75}{100} = 75\%$
5. $\frac{4}{20} = \frac{20}{100} = 20\%$
6. $\frac{3}{9} = \frac{33\frac{1}{3}}{100} = 33\frac{1}{3}\%$

Page 59: Decimals, Fractions, and Percents Review

1. $\frac{7}{10} = \frac{70}{100} = .70 = 70\%$
2. $\frac{1}{2} = \frac{50}{100} = .50 = 50\%$
3. $\frac{5}{8} = \frac{62.5}{100} = .625 = 62.5\%$
4. <
5. >
6. 12.5%
7. 25%
8. 75%
9. $\frac{2}{8}$; 25%
10. $\frac{3}{5}$; 60%

THE MEANING OF PERCENT

Page 60: Mixed Review

	Fraction	Decimal	Percent
1.	$\frac{3}{5}$	.60	60%
2.	$\frac{3}{4}$	.75	75%
3.	$\frac{1}{3}$	$.33\frac{1}{3}$	$33\frac{1}{3}\%$
4.	$\frac{1}{5}$	.20	20%
5.	$\frac{1}{2}$	.50	50%
6.	$\frac{7}{10}$	.70	70%
7.	$\frac{1}{8}$	$.12\frac{1}{2}$	$12\frac{1}{2}\%$
8.	$\frac{1}{4}$	.25	25%

Page 61: Change Fractions to Percents

1. $\frac{3}{4} = \frac{75}{100}$ 75 hundredths = 75%

2. $\frac{1}{3} = \frac{33\frac{1}{3}}{100}$ $33\frac{1}{3}$ hundredths = $33\frac{1}{3}\%$

3. $\frac{4}{5} = \frac{80}{100}$ 80 hundredths = 80%

4. $\frac{2}{25} = \frac{8}{100}$ 8 hundredths = 8%

5. $\frac{3}{8} = \frac{37\frac{1}{2}}{100}$ $37\frac{1}{2}$ hundredths = $37\frac{1}{2}\%$

6. $\frac{9}{10} = \frac{90}{100}$ 90 hundredths = 90%

7. $\frac{5}{6} = \frac{83\frac{1}{3}}{100}$ $83\frac{1}{3}$ hundredths = $83\frac{1}{3}\%$

8. $\frac{5}{8} = \frac{62\frac{1}{2}}{100}$ $62\frac{1}{2}$ hundredths = $62\frac{1}{2}\%$

9. $\frac{2}{3} = \frac{66\frac{2}{3}}{100}$ $66\frac{2}{3}$ hundredths = $66\frac{2}{3}\%$

10. $\frac{1}{6} = \frac{16\frac{2}{3}}{100}$ $16\frac{2}{3}$ hundredths = $16\frac{2}{3}\%$

Page 62: Using Division

	Divide	Change to a Percent
1.	$10\overline{)3.00}$ = .30	.30 = 30%
2.	$10\overline{)9.00}$ = .90	.90 = 90%
3.	$4\overline{)3.00}$ = .75	.75 = 75%
4.	$5\overline{)3.00}$ = .60	.60 = 60%
5.	$2\overline{)1.00}$ = .50	.50 = 50%

Page 63: Working with Remainders

	Divide	Change to a Percent
1.	$8\overline{)7.00}$ = $.87\frac{4}{8} = .87\frac{1}{2}$	$.87\frac{1}{2} = 87\frac{1}{2}\%$
2.	$8\overline{)3.00}$ = $.37\frac{4}{8} = .37\frac{1}{2}$	$.37\frac{1}{2} = 37\frac{1}{2}\%$
3.	$8\overline{)5.00}$ = $.62\frac{4}{8} = .62\frac{1}{2}$	$.62\frac{1}{2} = 62\frac{1}{2}\%$
4.	$9\overline{)1.00}$ = $.11\frac{1}{9}$	$.11\frac{1}{9} = 11\frac{1}{9}\%$
5.	$3\overline{)2.00}$ = $.66\frac{2}{3}$	$.66\frac{2}{3} = 66\frac{2}{3}\%$
6.	$6\overline{)5.00}$ = $.83\frac{2}{6} = .83\frac{1}{3}$	$.83\frac{1}{3} = 83\frac{1}{3}\%$

Page 64: Mixed Numbers to Percents

	Improper Fraction	Think Division	Write Percent
1.	$\frac{15}{4}$	$\begin{array}{r}3.75\\4\overline{)15.00}\end{array}$	375%
2.	$\frac{7}{3}$	$\begin{array}{r}2.33\frac{1}{3}\\3\overline{)7.00}\end{array}$	$233\frac{1}{3}\%$
3.	$\frac{9}{2}$	$\begin{array}{r}4.50\\2\overline{)9.00}\end{array}$	450%
4.	$\frac{23}{6}$	$\begin{array}{r}3.83\frac{2}{6} = 3.83\frac{1}{3}\\6\overline{)23.00}\end{array}$	$383\frac{1}{3}\%$
5.	$\frac{23}{8}$	$\begin{array}{r}2.87\frac{4}{8} = 2.87\frac{1}{2}\\8\overline{)23.00}\end{array}$	$287\frac{1}{2}\%$
6.	$\frac{25}{4}$	$\begin{array}{r}6.25\\4\overline{)25.00}\end{array}$	625%

Page 65: Use the Symbols

1. $>$
2. $<$
3. $<$
4. $=$
5. $>$
6. $<$
7. $>$
8. $<$
9. $>$
10. $>$
11. $<$
12. $>$
13. $<$
14. $>$
15. $>$
16. $=$

Page 66: More Fractions and Percents Review

1. $\frac{3}{25} = \frac{12}{100}$
12 hundredths = 12%

2. $\frac{1}{5} = \frac{20}{100}$
20 hundredths = 20%

3. $433\frac{1}{3}\%$

4. 262.5%

5. 20%

6. 37.5%

7. 60%

8. 80%

Page 67: Understanding Percents

1.	$\frac{2}{20} = \frac{1}{10}$	.10	10%
2.	$\frac{6}{24} = \frac{1}{4}$	.25	25%
3.	$\frac{9}{10}$	.90	90%
4.	$\frac{3}{4}$	.75	75%
5.	$\frac{8}{10} = \frac{4}{5}$	.80	80%
6.	$\frac{10}{30} = \frac{1}{3}$	$.33\frac{1}{3}$	$33\frac{1}{3}\%$

Page 68: Explain the Meaning

Answers should be similar to these:

1. All students were in the class.

2. The original price has been cut in half.

3. For every $1.00 on the bill, Allen paid $.15.

4. $\frac{1}{3}$ has been taken off the original price.

5. 35 parts out of 100 are wool and 65 parts out of 100 are cotton.

6. 9 out of 10 students passed the test.

7. Food that cost $1 last year costs $1.10 this year.

8. For every dollar's worth of goods sold, the commission is $.20.

Page 69: Percents of a Circle

1. 50%

2. $33\frac{1}{3}\%$

3. 25%

4. 20%

5. $12\frac{1}{2}\%$

6. 10%

Page 70: Circle Graphs

1. 20%
2. 30%
3. 25%
4. 25%
5. delivering papers
6. 35%
7. food
8. rent
9. utilities
10. utilities

Page 71: Practice Your Skills

1. XXXXX / XXXXX (first three of each row circled)
2. ●●○○○
3. 75%
4. $33\frac{1}{3}\%$
5. 30%
6. 70%
7. 20%
8. 80%
9. $66\frac{2}{3}\%$
10. $33\frac{1}{3}\%$

Page 72: Percent Review

1. 25%
2. 40%
3. $66\frac{2}{3}\%$
4. ■■■■□
5. ■□□□ / ■□□□ / ■□□□

	Fraction	Decimal	Percent
6.	$\frac{1}{2}$	.50	50%
7.	$\frac{1}{4}$	.25	25%
8.	$\frac{3}{10}$	.30	30%
9.	$\frac{3}{4}$	.75	75%
10.	$\frac{1}{5}$	.20	20%
11.	$\frac{3}{8}$	$.37\frac{1}{2}$	$37\frac{1}{2}\%$
12.	$\frac{1}{20}$	.05	5%
13.	$\frac{1}{100}$	.01	1%
14.	$\frac{1}{8}$	$.12\frac{1}{2}$	$12\frac{1}{2}\%$
15.	$\frac{1}{3}$	$.33\frac{1}{3}$	$33\frac{1}{3}\%$

Page 73: Real-Life Percents

1. 27
2. 50%
3. $\frac{1}{2}$
4. $66\frac{2}{3}\%$
5. $\frac{3}{4}$
6. 40%
7. 233%
8. 5

Page 74: Real-Life Percents Review

1. Answers will vary.
2. Answers will vary.
3. Answers will vary.
4. Answers will vary.
5. $\frac{1}{4}$
6. 20%
7. $\frac{3}{5}$
8. $66\frac{2}{3}\%$

Page 75: Cumulative Review

1. a) 16
   b) 40
2. a) $\frac{3}{5}$
   b) $33\frac{1}{3}\%$
3. a) $\frac{23}{50}$
   b) $1\frac{1}{20}$
   c) $3\frac{1}{2}$
   d) $1\frac{21}{50}$
4. a) 10%
   b) 90%
   c) 30%
5. a) 67%
   b) 50%
   c) 25%
   d) $33\frac{1}{3}\%$

Page 76: Posttest

1. 65%
2. 18
3. 20%
4. 40%
5. $\frac{24}{100} = \frac{6}{25}$
6. 139%
7. 260%
8. 85%
9. $62\frac{1}{2}\%$
10. 300%
11. $4\frac{1}{2}$
12. 15%
13. 80%
14. 600%
15. 2.5%
16. .07
17. 95%
18. 70%
19. 29%
20. $37\frac{1}{2}\%$

Posttest Evaluation Chart

If a student misses one or more problems in a skill area, assign a review of the practice pages for that skill.

Skill Area	Posttest Problem #	Skill Section	Review Page
Meaning of Percent	2, 8, 10	7–17	18
Fractions and Percent	3, 5, 9, 12, 18	19–34 61–65	35 66
Percents Greater than 100	7, 11, 14	36–40	41
Percents and Decimals	15, 16, 19	42–50	51
Decimals, Fractions, and Percents	1, 6	52–58	59
Percent Problem Solving	4, 13, 17, 20	67–73	74

PERCENT APPLICATIONS

Throughout this book students will learn to apply percents to real-life problem solving. To do this, they will learn more about number sentences, and they will learn to use a convenient memory device called the percent circle.

CONTENTS OF STUDENT TEXT	PAGE NUMBERS

Student Glossary

Acquainting students with definitions of key math terms and life-skills concepts will enhance their mastery of the materials. Below are words defined in the student text. A glossary is provided at the end of the student text and on page 243 of the *Teacher's Resource Guide and Answer Key*.

budget
commission
discount
down payment
earnings
equivalent
finance charge
interest
percent
premium
principal
reduction
salary
sales tax
savings account
simplify (reduce)
tax rate
tip
weekly allowance

STUDENT PAGE
7
Percent Problems

Number Sentences

After students have developed an understanding of the meaning of percent, they are ready to work with mathematical sentences.

Students will understand a **mathematical sentence** if you break it down into three important items: (1) **percent,** (2) **total,** and (3) **part.**

Discuss the three parts of a mathematical sentence.

EXAMPLE 50% of 6 = 3

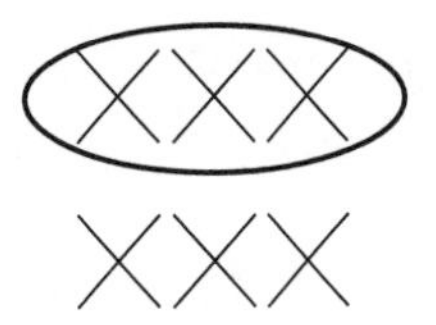

Percent: From the mathematical sentence 50% of 6 = 3, what percent is given? (Answer: *50%*)

Total: From the mathematical sentence 50% of 6 = 3, what represents the total? (Answer: *6*)

Part: From the mathematical sentence 50% of 6 = 3, what number represents the part being compared to the total number of Xs? (Answer: 3)

STUDENT PAGE
18
Find the Part

Use a Grid

To review finding the percent of a number, use a grid. You may place more numbers along the top and side.

X	$20	$55	$98	$75
15%	$3	$8.25		
25%		$13.75		
30%				
50%				

Alternate Activity

Using grocery store advertisements, newspapers, and magazines, have students create a chart of prices, percents, and discounts.

STUDENT PAGE
24
Find the Percent

Equal Ratios

Students often have difficulty with finding the percent when the total and part are given. Using a chart can simplify the process.

EXAMPLE

Discuss this problem: $n\%$ of 25 = 6

Show students that the problem requires a ratio of 6 to 25. Ask them to write a chart with a series of equal ratios that goes to 100.

Point out that you have two equal ratios known as a **proportion**.

$\frac{6}{25} = \frac{24}{100}$ or 24%

Equal Ratios	
6	25
12	50
18	75
24	100

← $\frac{24}{100}$ or 24%

STUDENT PAGE
37
Find the Total

Mentally Finding the Total

Present the class with problems for finding the total when the percent and part are given. Ask students to work many of these problems without using a pencil and paper. This activity will help students determine if the answer reasonable.

EXAMPLES

A 50% of n = 6 (percent: 50%, total: n, part: 6) Think: If 50% = 6, what is 100%? (Answer: 12)

B 25% of n = 15 Think: If 25% = 15, what is 100%? (Answer: 60)

C 10% of n = 4 Think: If 10% = 4, what is 100%? (Answer: 40)

STUDENT PAGE
44
Percent Problem Solving

Working Through Number Sentences

Have students arrange a series of numbers to make each number sentence true.

EXAMPLES

_____% of _____ is _____. 30 60 50

_____% of _____ is _____. 70 21 30

_____% of _____ is _____. 90 10 9

_____% of _____ is _____. 70 280 196

STUDENT PAGE
52
Percent Problem Solving

Use the Percent Circle

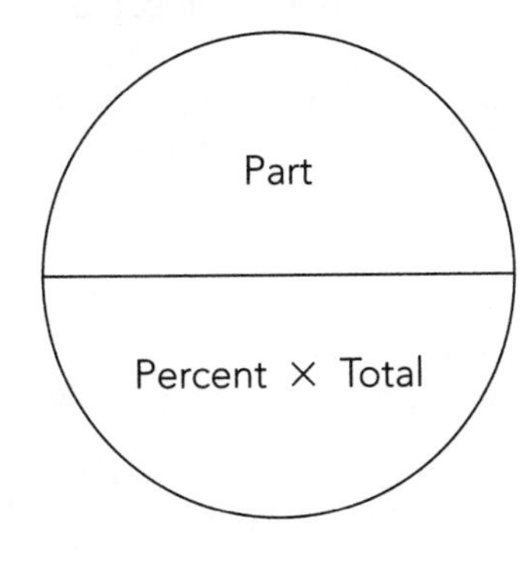

Ask students to bring in pictures from newspapers, magazines, or sales catalogs where percents are used in advertisements. Write a large percent circle on the blackboard and substitute what is known in the problem. Then solve the problem.

EXAMPLE A $96 jacket was marked off 25%. How much can you save by buying the jacket on sale?

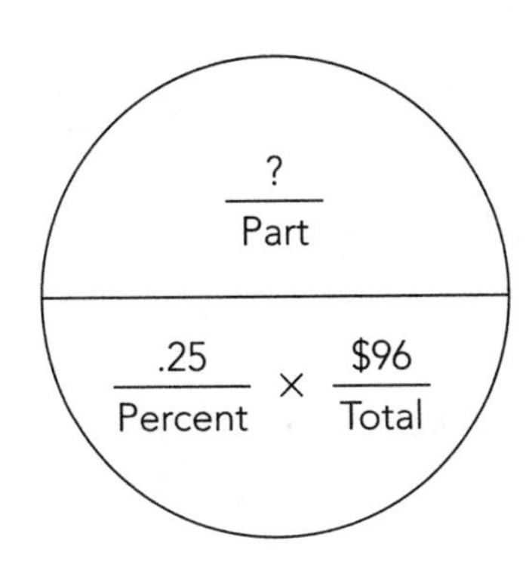

Using the percent circle, students can visually set up the problem. The part can be found by multiplying the percent as a decimal times the total.

PERCENT APPLICATIONS

STUDENT PAGE
54
Percent Problem Solving

Set Up a Proportion

Another approach to solving percent problems is to set up a proportion. Ask students to bring in pictures from newspapers, magazines, or sales catalogs where percents are used in advertisements. Make up problems and solve using the proportion method.

EXAMPLES

A Mr. Jones bought a jacket for 20% off the original price of $125. How much is saved buying the jacket on sale?

$\frac{\text{percent off}}{\text{total percent}} \quad \left\{\frac{20}{100} = \frac{n}{125}\right\} \quad \frac{\text{part}}{\text{total}}$ Answer: $25

B An $80 sweater was marked off $16. What percent was marked off?

$\frac{\text{percent off}}{\text{total percent}} \quad \left\{\frac{n}{100} = \frac{16}{80}\right\} \quad \frac{\text{part}}{\text{total}}$ Answer: 20%

C Nancy made a down payment of 20% on a television set. She paid $90 down. What was the price of the television set?

$\frac{\text{percent off}}{\text{total percent}} \quad \left\{\frac{20}{100} = \frac{90}{n}\right\} \quad \frac{\text{part}}{\text{total}}$ Answer: $450

STUDENT PAGE
54
Percent Problem Solving

Percent Box

An alternative to the **percent circle** is the **percent box.** While many students prefer the circle because of its simplicity, the percent box builds students' understanding of proportion. The percent box is always drawn the same way:

part			percent
total		100%	100%

You can teach students to fill in the box and then to cross multiply and divide.

EXAMPLE

Sarah had 30% of her typing done. If she had typed 6 chapters, how many chapters will she type altogether?

part	6	30	percent
total	n	100	100%

1 Cross multiply the filled-in numbers $6 \times 100 = 30 \times n$
$600 = 30 \times n$

2 Divide by the remaining number.
$600 \div 30 = n = 20$ chapters

Like the percent circle, the box can be used to find percent, part, and whole.

STUDENT PAGE
58
Life-Skills Math

Estimate Using 1%

Finding 1% of a number is easy. Just move the decimal point of the number being multiplied by 1% two places to the left.

EXAMPLES
1% of \$3.00 = \$.03
1% of \$4.60 = \$.046 or about \$.05
1% of \$384 = \$3.84

Make sure students can find and estimate using 1% of a number.

EXAMPLES

A 16% of \$3.45 = ______
Think: 1% of \$3.45 = \$.0345, which rounds to \$.03.
16 × .03 = \$.48

B 4% of \$16.95 = ______
Think: 1% of \$16.95 = \$.1695, which rounds to \$.17.
4 × .17 = \$.68

STUDENT PAGE
58
Life-Skills Math

Estimate Using 10%

Tell students to remember that 10% = $\frac{1}{10}$, and to find 10% of anything is the same as dividing it by 10. This can be accomplished by moving the decimal point one place to the left.

EXAMPLES

A What is 20% of 14?
Think: 10% of 14 = 1.4
20% is 2.8

B What is 20% of \$13.45?
Think: 10% of \$13.45 = \$1.345 or about \$1.35
20% of \$13.45 is about \$2.70

C What is 20% of \$3.98?
Think: 10% of \$3.98 = \$.398 or about \$.40
20% of \$3.98 is about \$.80

D What is 5% of \$14.98?
Think: 10% of \$14.98 = \$1.498 or about \$1.50
5% of \$14.98 is about \$.75

Patterns can be most helpful in solving everyday percent problems. Exercises like this can be repeated several times during the year.

STUDENT PAGE
58
Life-Skills Math

Mental Computation

Explain to students that a great many percent problems can be solved without pencil and paper. Use common percent equivalents to mentally find the answer.

EXAMPLES

What is 60% of 14?
Think: 50% of 14 = 7
10% of 14 = 1.4
60% of 14 = 7 + 1.4 = 8.4

Use real-life examples. For instance, when you figure a tip at a restaurant, you often round first, then mentally find the tip.

The bill is $19.47. What is the 15% tip?
Round $19.47 to $20.
Think: 10% of $20 = $2
5% of $20 = $1
$2 + $1 = $3
The tip is $3.

STUDENT PAGE
58
Life-Skills Math

Watch for Patterns

Percent, as in all mathematics, is a subject in which students can learn to look for patterns. Explain to students that a great many percent problems can be compared to patterns that make up easier problems.

EXAMPLES

A What is 14% of 50?
Would this be the same as 50% of 14?
(Answer: Yes.)
Switch the factors and come up with a quick answer.
(Answer: $\frac{1}{2} \times 14 = 7$)

B What is 40% of 25?
Would this be the same as 25% of 40?
Switch the factors and the rest is easy.
(Answer: $\frac{1}{4} \times 40 = 10$)

C What is 12% of 75?
Would this be the same as 75% of 12?
(Answer: $\frac{3}{4} \times 12 = 9$)

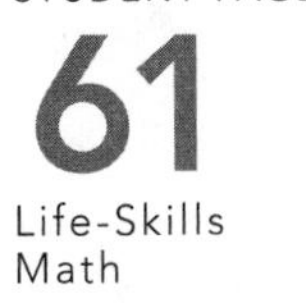

Depreciation

A car depreciates in value each year during its lifetime. Each year the resale value of a car is less. Have students complete the tables according to the schedules shown.

Cost of Car	Year	Percent of Depreciation	Amount of Depreciation	Resale Value
$12,585	1	20%	$2,517 a	$10,068 b

Cost of Car	Year	Percent of Depreciation	Amount of Depreciation	Resale Value
$26,700	1	23%	______ a	______ b

Cost of Car	Year	Percent of Depreciation	Amount of Depreciation	Resale Value
$15,950	1	22%	______ a	______ b

How Much Will You Save?

Compare and see how much you can save on annual credit card interest.

Interest Savings Chart					
		Annual interest Paid on an Average Daily Balance of:			
Credit Card	Interest Rate	$1,500	$2,000	$2,500	$5,000
Bankcard "A"	14.9%	$223.50	$298.00	$372.50	$745.00
Bankcard "B"	17.5%	$262.50	$350.00	$437.50	$875.00
Bankcard "C"	19.8%	$297.00	$396.00	$495.00	$990.00
Bankcard "D"	21.6%	$324.00	$432.00	$540.00	$1,080.00

SAMPLE QUESTIONS

What is the difference in interest rates between Bankcard A and Bankcard C? (Answer: 4.9%)

How much do you save by using Bankcard A over Bankcard D on a balance of $2,000? (Answer: $134.00)

What is the interest rate for Bankcard B? (Answer: 17.5%)

Page 4: Pretest

1. 50%
2. 20
3. 27
4. 45
5. 308
6. 13.75
7. 63
8. 40
9. 26%
10. 75%
11. $37\frac{1}{2}\%$
12. $66\frac{2}{3}\%$
13. 150%
14. 60
15. 150
16. $78
17. 85%
18. $1,900
19. 15%
20. 21 students

Pretest Evaluation Chart

If a student misses any problems in a skill area, assign the practice pages for that skill. However, students may need to complete all practice pages to reinforce areas of weakness.

Skill Area	Pretest Problem #	Skill Section	Review Page
Percent Problems	All	7–11	12
Find the Part	2, 4, 5, 6, 7, 8	13–21	22
Find the Percent	1, 9, 10, 11, 12, 13	7–11 23–35	12 36
Find the Total	3, 14, 15	7–11 37–42	12 43
Percent Problem Solving	16, 17, 18, 19, 20	44–55	56
Life-Skills Math	All	57–72	73, 74

Page 7: Percent Applications

1. $5.00
2. $1.50
3. 30%

Page 8: Percent Readiness

	Part Circled	Total—100% of Xs	Percent
1.	2	4	50%
2.	9	12	75%
3.	8	20	40%
4.	3	9	$33\frac{1}{3}\%$
5.	2	10	20%
6.	5	8	$62\frac{1}{2}\%$

Page 9: Understanding Number Sentences

	Diagram	Part Circled	Total—100% of Xs	Percent
1.	(XX)XX XXXX	2	8	25%
2.	(XXX)X	3	4	75%
3.	(XXX)XX XXXXX XXXXX	3	15	20%
4.	(X)XX	1	3	$33\frac{1}{3}\%$
5.	(XXX)XX XXXXX	3	10	30%
6.	(XX)XX XXXX XXXX XXXX	2	16	$12\frac{1}{2}\%$

Page 10: Showing Relationships

1. 10
2. 4
3. 40%
4. 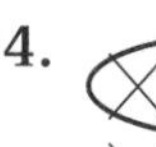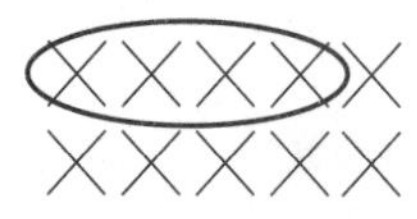

	Total	Part	Percent
5.	70	7	10%
6.	50	30	60%
7.	10	3	30%
8.	9	6	$66\frac{2}{3}\%$
9.	300	100	$33\frac{1}{3}\%$
10.	90	45	50%
11.	12	10	$83\frac{1}{3}\%$
12.	60	42	70%

Page 11: Identify the Part, Total, and Percent

	Part	Total	Percent
1.	16	80	20%
2.	21	30	70%
3.	5.7	38	15%
4.	192	96	200%
5.	8.04	67	12%
6.	63	70	90%
7.	267	89	300%
8.	28	70	40%

Page 12: Percent Problems Review

1. a) Part: 3
   b) Total: 12
   c) Percent: 25
2. a) Part: 4
   b) Total: 10
   c) Percent: 40
3. a) Part: 4
   b) Total: 8
   c) Percent: 50
4. a) Part: 2
   b) Total: 6
   c) Percent: $33\frac{1}{3}$
5. a) Part: 90
   b) Total: 45
   c) Percent: 200
6. a) Part: 10.6
   b) Total: 42.4
   c) Percent: 25
7. a) Part: 259
   b) Total: 74
   c) Percent: 350
8. a) Part: 27
   b) Total: 90
   c) Percent: 30

Page 13: Find the Percent of a Number

1. $.30 \times 70 = n$
   $21 = n$
2. $.80 \times 50 = n$
   $40 = n$

Page 14: Change Percents to Decimals

1. $.19 \times 17 = n$
   $3.23 = n$
2. $.95 \times 960 = n$
   $912 = n$
3. $.08 \times 23 = n$
   $1.84 = n$
4. $.82 \times 194 = n$
   $159.08 = n$
5. $.48 \times 39 = n$
   $18.72 = n$
6. $.15 \times 598 = n$
   $89.7 = n$

Page 15: Small and Large Percents of a Number

A. $1.27 \times 89 = n$
   $113.03 = n$
B. $.004 \times 89 = n$
   $.356 = n$
1. $2 \times 348 = n$
   $696 = n$
2. $1.12 \times 95 = n$
   $106.4 = n$
3. $3.45 \times 70 = n$
   $241.5 = n$
4. $.008 \times 32 = n$
   $.256 = n$
5. $.003 \times 285 = n$
   $.855 = n$
6. $.025 \times 321 = n$
   $8.025 = n$

Page 16: Change Percents to Fractions

1. $\frac{1}{2} \times 80 = n$
   $40 = n$
2. $\frac{1}{5} \times 40 = n$
   $8 = n$
3. $\frac{1}{10} \times 60 = n$
   $6 = n$
4. $\frac{1}{4} \times 32 = n$
   $8 = n$

Page 17: Using Fractions

1. $\frac{1}{2} \times 20 = n$
   $10 = n$
2. $\frac{1}{10} \times 90 = n$
   $9 = n$
3. $\frac{1}{5} \times 50 = n$
   $10 = n$
4. $\frac{3}{4} \times 80 = n$
   $60 = n$
5. $\frac{1}{4} \times 120 = n$
   $30 = n$
6. $\frac{1}{2} \times 180 = n$
   $90 = n$

Page 18: Using Common Equivalents

1. $6 = n$
2. $\frac{1}{10} \times \$1{,}450 = n$
   $\$145 = n$
3. $\frac{1}{3} \times 45 = n$
   $15 = n$
4. $\frac{1}{5} \times 25 = n$
   $5 = n$
5. $\frac{1}{8} \times 720 = n$
   $90 = n$
6. $\frac{3}{4} \times 88 = n$
   $66 = n$

Page 19: Mixed Practice

Number sentences may vary.

1. $\frac{1}{2} \times 90 = n$
   $45 = n$
2. $.39 \times 145 = n$
   $56.55 = n$
3. $\frac{1}{4} \times 36 = n$
   $9 = n$
4. $.05 \times 4{,}508 = n$
   $225.4 = n$
5. $\frac{1}{3} \times 15 = n$
   $5 = n$
6. $1.75 \times 36 = n$
   $63 = n$
7. $.003 \times 910 = n$
   $27.3 = n$
8. $\frac{1}{10} \times 300 = n$
   $30 = n$
9. $.80 \times \$58.65 = n$
   $\$46.92 = n$
10. $\frac{1}{5} \times 35 = n$
    $7 = n$

Page 20: Write a Number Sentence

1. a) 15%
   b) $495
   c) n
2. 15% of $\$495 = n$
   $.15 \times \$495 = n$
   $\$74.25 = n$
3. a) 25%
   b) $840
   c) n
4. 25% of $\$840 = n$
   $\frac{1}{4} \times \$840 = n$
   $\$210 = n$

Page 21: Solve the Word Problems

1. 25% of $\$52 = n$
   $\$13 = n$
2. 5% of $\$94.60 = n$
   $\$4.73 = n$
3. 8% of $600 = n$
   $48 = n$
4. 75% of $140 = n$
   $105 = n$
5. 12% of $\$947 = n$
   $\$113.64 = n$
6. 30% of $\$435 = n$
   $\$130.50 = n$
7. 50% of $\$300 = n$
   $\$150 = n$
8. 10% of $\$169{,}500 = n$
   $\$16{,}950 = n$

Page 22: Find the Part Review

1. $.2 \times 18 = 3.6$; $3.6 = n$
2. $.95 \times 39 = 37.05$; $37.05 = n$
3. $.07 \times 25 = 1.75$; $1.75 = n$
4. $.5 \times 84 = 42$; $42 = n$
5. $.2 \times 330 = 66$; $66 is saved
6. $.125 \times 440 = 110$; 110 runners finished
7. $45.59 \times .08 = \$3.65$ sales tax
8. $.4 \times 750 = 300$; $300 saved

Page 23: Compare the Numbers

	Picture	Fraction	Decimal	Percent
1.		$\frac{3}{12} = \frac{1}{4} = \frac{25}{100}$	.25	25%
2.		$\frac{4}{20} = \frac{20}{100}$	.20	20%
3.		$\frac{7}{10} = \frac{70}{100}$	.70	70%
4.		$\frac{4}{8} = \frac{1}{2} = \frac{50}{100}$	.50	50%
5.		$\frac{37}{50} = \frac{74}{100}$	.74	74%

Page 24: Find the Percent

1. $\frac{8}{10} = \frac{80}{100} = 80\%$
2. $\frac{7}{25} = \frac{28}{100} = 28\%$
3. $\frac{18}{20} = \frac{90}{100} = 90\%$
4. $\frac{3}{4} = \frac{75}{100} = 75\%$

Page 25: Simplify the Fraction

1. $\frac{5}{25} = \frac{20}{100} = 20\%$
2. $\frac{15}{30} = \frac{1}{2} = \frac{50}{100} = 50\%$
3. $\frac{4}{40} = \frac{1}{10} = \frac{10}{100} = 10\%$
4. $\frac{24}{32} = \frac{3}{4} = \frac{75}{100} = 75\%$

Page 26: More Practice Finding the Percent

1. $\frac{8}{40} = \frac{1}{5} = \frac{20}{100} = 20\%$
2. $\frac{10}{40} = \frac{1}{4} = \frac{25}{100} = 25\%$
3. $\frac{3}{20} = \frac{15}{100} = 15\%$
4. $\frac{3}{10} = \frac{30}{100} = 30\%$
5. $\frac{4}{50} = \frac{8}{100} = 8\%$
6. $\frac{32}{40} = \frac{4}{5} = \frac{80}{100} = 80\%$
7. $\frac{12}{16} = \frac{3}{4} = \frac{75}{100} = 75\%$
8. $\frac{7}{70} = \frac{1}{10} = \frac{10}{100} = 10\%$

Page 27: Use Division

1. $\frac{5}{8} = 62\frac{1}{2}\%$
2. $\frac{15}{48} = \frac{5}{16} = 31\frac{1}{4}\%$
3. $\frac{5}{6} = 83\frac{1}{3}\%$
4. $\frac{30}{32} = \frac{15}{16} = 93\frac{3}{4}\%$

Page 28: Changing Fractions to Decimals

1. $\frac{3}{11} = 27\frac{3}{11}\%$
2. $\frac{7}{21} = \frac{1}{3} = 33\frac{1}{3}\%$
3. $\frac{3}{21} = \frac{1}{7} = 14\frac{2}{7}\%$
4. $\frac{21}{24} = \frac{7}{8} = 87\frac{1}{2}\%$
5. $\frac{2}{18} = \frac{1}{9} = 11\frac{1}{9}\%$
6. $\frac{14}{21} = \frac{2}{3} = 66\frac{2}{3}\%$

Page 29: What Is the Percent?

1. $\frac{6}{24} = 25\%$
2. $\frac{7}{35} = 20\%$
3. $\frac{7}{21} = 33\frac{1}{3}\%$
4. $\frac{25}{50} = 50\%$
5. $\frac{16}{24} = 66\frac{2}{3}\%$
6. $\frac{15}{40} = 37\frac{1}{2}\%$
7. $\frac{5}{7} = 71\frac{3}{7}\%$
8. $\frac{32}{40} = 80\%$

Page 30: Fractions and Percents Greater Than 100%

Fraction	Percent
1. $\frac{13}{10} = \frac{130}{100}$	130%
2. $\frac{9}{5} = \frac{180}{100}$	180%
3. $\frac{6}{4} = \frac{150}{100}$	150%
4. $\frac{7}{2} = \frac{350}{100}$	350%
5. $\frac{32}{20} = \frac{160}{100}$	160%

Page 31: Fractions Larger Than 1

1. $\frac{5}{4} = \frac{125}{100} = 125\%$
2. $\frac{15}{10} = \frac{150}{100} = 150\%$
3. $\frac{9}{2} = \frac{450}{100} = 450\%$
4. $\frac{50}{20} = \frac{250}{100} = 250\%$
5. $\frac{21}{6} = \frac{7}{2} = \frac{350}{100} = 350\%$
6. $\frac{35}{20} = \frac{175}{100} = 175\%$
7. $\frac{40}{25} = \frac{160}{100} = 160\%$
8. $\frac{22}{4} = \frac{550}{100} = 550\%$

Page 32: Decimals and Percents Greater Than 100%

1. $\frac{9}{2} = 450\%$
2. $\frac{10}{3} = 333\frac{1}{3}\%$
3. $\frac{8}{5} = 160\%$
4. $\frac{15}{7} = 214\frac{2}{7}\%$
5. $\frac{16}{6} = 266\frac{2}{3}\%$
6. $\frac{13}{4} = 325\%$

Page 33: Practice Your Skills

1. $\frac{9}{20} = 45\%$
2. $\frac{17}{85} = 20\%$
3. $\frac{27}{9} = 300\%$
4. $\frac{5}{2} = 250\%$
5. $\frac{15}{60} = 25\%$
6. $\frac{8}{7} = 114\frac{2}{7}\%$
7. $\frac{22}{5} = 440\%$
8. $\frac{20}{40} = 50\%$
9. $\frac{18}{4} = 450\%$
10. $\frac{12}{18} = 66\frac{2}{3}\%$

Page 34: Solve the Word Problems

1. a) $n\%$
   b) 20
   c) 15
   d) $\frac{15}{20} = 75\%$
2. a) $n\%$
   b) 30
   c) 6
   d) $\frac{6}{30} = 20\%$
3. a) $n\%$
   b) 16
   c) 4
   d) $\frac{4}{16} = 25\%$
4. a) $n\%$
   b) 25
   c) 8
   d) $\frac{8}{25} = 32\%$

Page 35: More Word Problems

1. $n\%$ of 45 = 15
   $\frac{15}{45} = 33\frac{1}{3}\%$
2. $n\%$ of 24 = 6
   $\frac{6}{24} = 25\%$
3. $n\%$ of 400 = 40
   $\frac{40}{400} = 10\%$
4. $n\%$ of \$60 = \$30
   $\frac{30}{60} = 50\%$
5. $n\%$ of 6 = 4
   $\frac{4}{6} = 66\frac{2}{3}\%$
6. $n\%$ of \$450 = \$90
   $\frac{90}{450} = 20\%$
7. $n\%$ of 30 = 10
   $\frac{10}{30} = 33\frac{1}{3}\%$
8. $n\% \times 15 = 12$
   $\frac{12}{15} = 80\%$

Page 36: Find the Percent Review

1. 66.67%
2. 46%
3. 10%
4. 75%
5. 300%
6. 500%
7. 400%
8. 350%

Page 37: What Is the Total?

1. $.30 \times n = 21$
   $n = 70$
   30% of 70 is 21.
2. $.40 \times n = 32$
   $n = 80$
   40% of 80 is 32.
3. $.80 \times n = 16$
   $n = 20$
   80% of 20 is 16.
4. $.68 \times n = 153$
   $n = 225$
   68% of 225 is 153.

Page 38: Find the Total

1. $.75 \times n = 225$
   $n = 300$
2. $.04 \times n = 34$
   $n = 850$
3. $.25 \times n = 15$
   $n = 60$
4. $.15 \times n = \$4.65$
   $n = \$31$

Page 39: Find the Total When the Part Is Given

1. $.03 \times n = 4.23$
   $n = 141$
2. $.60 \times n = 42$
   $n = 70$
3. $.06 \times n = 18$
   $n = 300$
4. $.10 \times n = 6$
   $n = 60$
5. $.25 \times n = 40$
   $n = 160$
6. $.80 \times n = 5$
   $n = 5$
7. $.15 \times n = 30$
   $n = 200$
8. $.90 \times n = 45$
   $n = 50$

Page 40: Percent Problem Solving

1. a) 20%
   b) n
   c) $8.00
   d) 20% of $n = \$8.00$
   e) $40.00
2. a) 25%
   b) n
   c) $150.00
   d) 25% of $n = \$150.00$
   e) $600.00

Page 41: Solve the Word Problems

1. 6% of $n = \$30.00$
   $500.00
2. 15% of $n = \$1.80$
   $12.00
3. 20% of $n = \$156.00$
   $780.00
4. 10% of $n = 3$
   30
5. 40% of $n = \$200.00$
   $500.00
6. 15% of $n = \$39.00$
   $260.00
7. 25% of $n = \$9.00$
   $39.00
8. 20% of $n = 49$
   245

Page 42: Practice Your Skills

1. 85
2. 67.5
3. 70
4. 50
5. $190.63
6. $18.67

Page 43: Find the Total Review

1. 60
2. 300
3. 498
4. 137.5
5. $175
6. $1,170
7. 250
8. 20

Page 44: Identify the Facts

1. 75% of 12 = 9
2. 25% of 80 = 20
3. 4% of 75 = 3
4. 10% of 30 = 3
5. 30% of 6 = 1.8
6. 45% of 20 = 9

Page 45: Learn the Percent Circle

1.

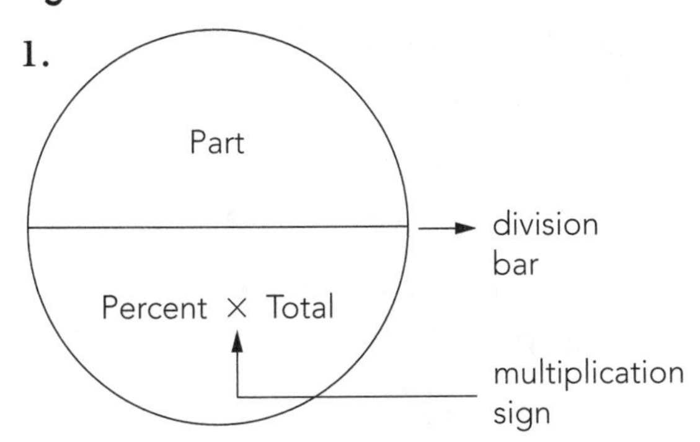

2.

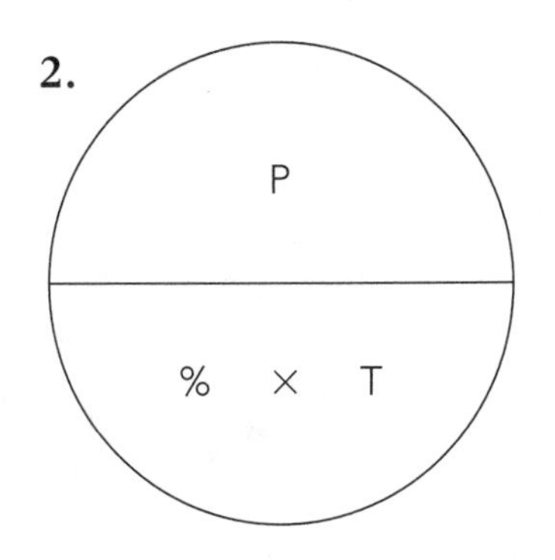

Page 46: Find the Part

1. a) P
   b) multiply
   c) 40% × 90 = 36
2. 15% × 75 = 11.25
3. 25% × 80 = 20
4. 30% × $45 = $13.50

Page 47: Find the Percent

1. a) %
   b) divide
   c) $\frac{7}{28} = \frac{1}{4} = 25\%$
2. $\frac{13}{65} = \frac{1}{5} = 20\%$
3. $\frac{8}{80} = \frac{1}{10} = 10\%$
4. $\frac{7}{175} = \frac{1}{25} = 4\%$

Page 48: Find the Total

1. a) T
   b) divide
   c) $\frac{7}{.20}$ $.20\overline{)7.00}$ = 35.
   7 is 20% of 35
2. $\frac{16}{.40}$ $.40\overline{)16.00}$ = 40.
   16 is 40% of 40
3. $\frac{35}{.70}$ $.70\overline{)35.00}$ = 50.
   35 is 70% of 50
4. $\frac{9}{.20}$ $.20\overline{)9.00}$ = 45.
   9 is 20% of 45

Page 49: Use the Circle

1. a) percent
   b) divide
   c) 16 ÷ 48 = n
   d) $n = 33\frac{1}{3}\%$
2. a) part
   b) multiply
   c) .75 × 60 = n
   d) n = 45
3. a) total
   b) divide
   c) 17 ÷ .05 = 340
   d) 340

Page 50: Problem-Solving Readiness

1. $\frac{1}{3} \times 45 = 15$
2. $15 \div 45 = 33\frac{1}{3}\%$
3. 12 ÷ .05 = 240
4. 16%
5. 2.3257
6. 26.25
7. 250%
8. 286
9. 136.5
10. 170

PERCENT APPLICATIONS

Page 51: Identify the Facts

A. 60%
C. 180
D. 60% of 300 = 180

1. 60% of \$90 = \$54

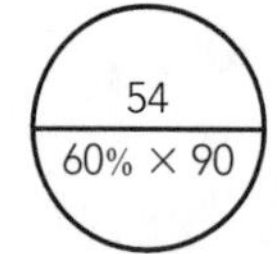

2. 15% of \$80 = \$12

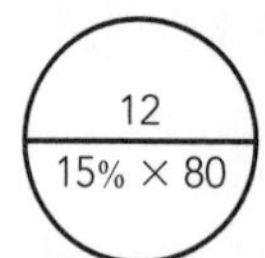

3. 25% of 180 = 45

45
25% × 180

4. 10% of \$350 = \$35

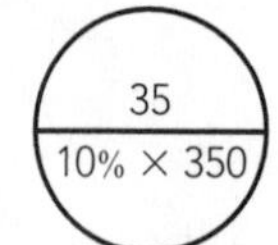

5. 8% of 50 = 4

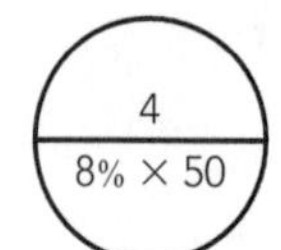

6. 80% of 150 = 120

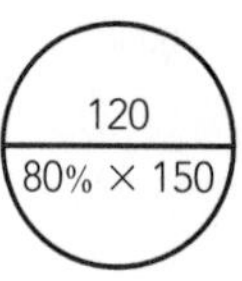

Page 52: Mixed Practice

1. 30% of \$20 = n; 6.00
2. n% of \$40.00 = 6.00; 15%
3. $33\frac{1}{3}$% of \$600 = n; 198.78
4. n% of 5 = 3; 60%
5. 75% of 8,000 = n; 6,000
6. 18% of n = \$27.00; 150.00
7. 10% of n = \$400.00; 4,000.00
8. 6% of \$45.00 = n; 2.70

Page 53: Problem Solving

1. 40% × n = \$34.00
The original price was \$85.00.
2. 6% × \$16,450
The sales tax was \$987.00
3. n% × \$4.50 = \$.90
The tip was 20% of the fare.
4. 8% × n = \$2,240
Her salary was \$28,000 before her raise.
5. $12\frac{1}{2}$% × 720 = n
She sold 90 candy bars.
6. 2.5% × \$840 = n
She earns \$21.00 interest in one year.
7. 5% × n = \$136.00
His total monthly earnings are \$2,720.
8. n% × 54 = 18
$33\frac{1}{3}$% times at bat were hits.

Page 54: Think It Through

1. b
2. c
3. a
4. f
5. d
6. e

Page 55: Practice Helps

1. 28
2. 75
3. 10%
4. 25%
5. 31.25
6. 35

Page 56: Percent Problem-Solving Review

1. 36
2. 22.5%
3. 80
4. \$182.00
5. 28
6. 20%
7. 250
8. \$38.50

Page 57: Round to the Nearest Cent

1. $8.85
2. $15.92
3. $37.86
4. $230.15
5. $9.66
6. $40.50
7. $6.71
8. $7.01
9. $1.18
10. $68.47

Page 58: Percent Off

1. 25% of $108 $= n$
   27 $= n$
   The amount of savings is $22.
2. 10% of $899.95 $= n$
   89.995 $= n$
   The amount of savings is $90.00.
3. 30% of $225 $= n$
   67.950 $= n$
   The amount of savings is $67.50.
4. 40% of $55.50 $= n$
   22.20 $= n$
   The amount of savings is $22.20.
5. 15% of $499.90 $= n$
   74.985 $= n$
   The amount of savings is $74.99.
6. 50% of $36.50 $= n$
   18.25 $= n$
   The amount of savings is $18.25.

Page 59: Discounts

	Amount of Discount	Sale Price
1.	a) $17.80	b) $71.20
2.	a) $5.10	b) $10.20
3.	a) $8.00	b) $24.00
4.	a) $62.80	b) $62.80
5.	a) $70.20	b) $163.80

Page 60: Find the Sale Price

1. a) $7.20
   b) $28.79
2. a) $199.75
   b) $599.25
3. a) $59.93
   b) $39.96
4. a) $3.89
   b) $22.06

Page 61: Discount Practice

1. a) $11.29
   b) $33.87
2. a) $65.00
   b) $260.00
3. a) $16.53
   b) $38.57
4. a) $898
   b) $17,062

	Amount of Discount	Sale Price
5.	a) $1.15	b) $1.15
6.	a) $11.84	b) $47.36
7.	a) $67.25	b) $201.75
8.	a) $1.86	b) $10.54
9.	a) $5.38	b) $21.50
10.	a) $.26	b) $2.34

Page 62: Sales Tax

Sales Tax

1. $.98
2. $1.27
3. $.50
4. $27.28
5. $11.51

Page 63: Find the Total Price

1. a) 22.95
   b) 29.97
   c) 52.92
   d) 3.18
   e) $56.10
2. a) 16.25
   b) 130.00
   c) 146.25
   d) 7.31
   e) $153.56
3. a) 225.00
   b) 225.00
   c) 16.88
   d) $241.88
4. a) 29.97
   b) 65.00
   c) 22.95
   d) 117.92
   e) 9.73
   f) $127.65

Page 64: Sales Tax Practice

1. a) $1.08
   b) $19.03
2. a) $.36
   b) $9.35
3. a) $6.29
   b) $132.08
4. a) $3.30
   b) $69.28

	Sales Tax	Total Purchase Price
5.	a) $2.93	b) $51.72
6.	a) $.24	b) $6.12
7.	a) $.68	b) $12.81
8.	a) $4.65	b) $71.04
9.	a) $1.20	b) $25.15
10.	a) $.25	b) $8.51

Page 65: Simple Interest in Savings

1. $7.50
2. $48.00
3. $11.00
4. $22.80
5. $27.00

Page 66: Borrowing Money and Paying Interest

	Interest	Amount to Be Repaid
1.	a) $63.00	b) $763.00
2.	a) $75.00	b) $325.00
3.	a) $1,200	b) $16,200
4.	a) $2,700	b) $10,200
5.	a) $560	b) $2,560

Page 67: Commission

1. 50% of $250 = n; 125.00
2. 15% of $82 = n; 12.30
3. 6% of $19,580 = n; 1,174.80
4. 20% of $5,960 = n; 1,192

Page 68: Commission Applications

1. 8% of $93,600 = n
   .08 × $93,600 = n
   $7,488 = n
2. n% of $1,590 = $79.50
   n = $79.50 ÷ $1,590
   n = 5%
3. n% of $350 = $175
   n = $175 ÷ $350
   n = 50%
4. 5% of n = $5,270
   n = $5,270 ÷ .05
   n = $105,400
5. 10% of $2,850 = n
   .10 × $2,850 = n
   $285 = n
6. 25% of $6,480 = n
   .25 × $6,480 = n
   $1,620 = n
7. 8% of $10,942 = n
   .08 × $10,942 = n
   $875.36 = n
8. 20% of $850 = n
   .20 × $850 = n
   $170 = n

Page 69: Practice with Commissions

1. 6% of $75,950 = n
   .06 × $75,950 = n
   $4,557 = n
2. 25% of n = $240
   n = $240 ÷ .25
   n = $960
3. n% of $1,200 = $240
   n = $240 ÷ $1,200
   n = 20%

4. 6% of \$95,292 = n
   .06 × \$95,292 = n
   \$5,717.52 = n

	Business	Rate	Total Sales	Commission
5.	Real Estate	4%	\$75,300	\$3,012
6.	Tile and Carpeting	20%	\$475.00	\$95.00
7.	Appliances	8%	\$300.00	\$24.00
8.	Insurance	5%	\$890.00	\$44.50
9.	Automobile	10%	\$15,700	\$1,570
10.	Clothing	12%	\$800.00	\$96.00

Page 70: Percents and Budgets

1. \$220
2. \$1,100
3. \$550
4. \$300
5. \$180
6. \$480
7. \$240

Page 71: Percent of Increase

1. 25%
2. 30%
3. 10%
4. 44%

Page 72: Percent of Decrease

1. 10%
2. 30%
3. 8%
4. 40%

Page 73: Life-Skills Math Review

1. a) \$1.79
   b) \$38.08
2. a) \$24.14
   b) \$582.01
3. \$29.25
4. a) \$154.00
   b) \$704.00
5. \$9,093.00
6. \$227.95
7. 33%
8. 29%

Page 74: Percent Application Review

1. \$4.97
2. 30%
3. \$24
4. a) 32
   b) 15%
   c) 120
5. a) \$3.99
   b) \$83.79
6. \$5,540
7. 25%
8. a) \$5.78
   b) \$32.74

Page 75: Cumulative Review

1. 22
2. 15%
3. 75%
4. 360
5. 7%
6. 20%
7. 45%
8. \$43.20
9. \$15.20
10. \$28.09

Page 76: Posttest

1. 13.5
2. 5
3. 21
4. $66\frac{2}{3}\%$
5. 64
6. 50%
7. $50
8. 160
9. 8
10. 75%
11. 95%
12. $150
13. 2%
14. 70
15. 64
16. $9.45
17. 250%
18. 200
19. 6%
20. 272

Posttest Evaluation Chart

If a student misses one or more problems in a skill area, assign a review of the practice pages for that skill.

Skill Area	Posttest Problem #	Skill Section	Review Page
Percent Problems	All	7–11	12
Find the Part	1, 5, 9, 14, 18, 20	13–21	22
Find the Percent	6, 10, 11, 13, 17	7–11 23–35	12 36
Find the Total	2, 8, 15	7–11 37–42	12 43
Percent Problem Solving	3, 4, 7, 12, 16, 19	44–55	56
Life-Skills Math	All	57–72	73, 74

Word	Definition	Student Book
A		
acre	an area of land	Ratio & Proportion
addition	to combine numbers and find a total	Whole Numbers: Addition & Subtraction
appointment	to meet someone at a specific time and place	Fractions: Addition & Subtraction
attendance	the number of people at an event	The Meaning of Percent Whole Numbers: Addition & Subtraction
average	the sum of numbers divided by the total amount of numbers—another name for *mean*	Decimals: Multiplication & Division Fractions: Multiplication & Division Ratio & Proportion
B		
batch	the amount made while baking	Fractions: Multiplication & Division
brand	a name identifying a product	Decimals: Multiplication & Division
bring down	to move the decimal point to the answer line	Decimals: Addition & Subtraction
budget	a plan for spending money	Decimals: Addition & Subtraction The Meaning of Percent Percent Applications
C		
centimeter	a unit of measure in the metric system equal to 100th of a meter	Fractions: Multiplication & Division
check	to make sure that a problem has been completed correctly	Whole Numbers: Addition & Subtraction
checking account	a bank account used to pay bills	Whole Numbers: Addition & Subtraction Decimals: Addition & Subtraction
circle	to draw a line around something	Decimals: Addition & Subtraction
clockwise	to turn in the direction that the hands of a clock move	Whole Numbers: Multiplication & Division
combined earnings	total earnings of more than one income	Decimals: Addition & Subtraction
commission	money paid to a sales person for his or her services, or money paid to a sales person based on the amount of items sold	The Meaning of Percent
comparison	to determine if two numbers are similar	Fractions: Addition & Subtraction
corresponding position	having the same relationship	Ratio & Proportion

Word	Definition	Student Book
coupons	a printed form or advertisement that can be used to reduce the price of an item	Decimals: Addition & Subtraction
cross product	when the numerator of one fraction and the denominator of another fraction are multiplied and vice versa	Ratio & Proportion

D

Word	Definition	Student Book
decimal place	the position of a number to the right of the decimal point	Decimals: Multiplication & Division
denominator	the bottom part of a fraction	Fractions: The Meaning of Fractions Fractions: Addition & Subtraction Fractions: Multiplication & Division Ratio & Proportion The Meaning of Percent Percent Applications
deposit	money put in a bank, or to put money in a bank	Decimals: Addition & Subtraction
difference	the answer to a subtraction problem	Whole Numbers: Addition & Subtraction
digit	one of the ten number symbols: 0, 1, 2, 3, 4, 5, 6, 7, 8, and 9	Whole Numbers: Addition & Subtraction Decimals: Addition & Subtraction
discount	the reduced cost of an item	Fractions: Multiplication & Division The Meaning of Percent Percent Applications
discount coupon	a printed form or advertisement that can be used to reduce the price of an item	Decimals: Addition & Subtraction
dividend	the number being divided	Whole Numbers: Multiplication & Division Decimals: Multiplication & Division
divisible	a number that can be divided, usually with no remainder	Fractions: The Meaning of Fractions
division	separating a number into equal parts	Whole Numbers: Multiplication & Division
divisor	the number doing the dividing; the number on the outside of the division box	Whole Numbers: Multiplication & Division Decimals: Multiplication & Division
down payment	a partial payment made at the time of purchase, with the balance due later	Decimals: Multiplication & Division Percent Applications

E

Word	Definition	Student Book
earnings	salary, wages, or income	Fractions: Multiplication & Division The Meaning of Percent Percent Applications
equivalent	two numbers that have the same value	The Meaning of Percent Percent Applications

Word	Definition	Student Book
equivalent fraction	fractions that have the same value	Fractions: The Meaning of Fractions
estimate	an approximate answer	Whole Numbers: Multiplication & Division
expanded form	to write in the long form	Whole Numbers: Addition & Subtraction Decimals: Addition & Subtraction

F

Word	Definition	Student Book
factor	pairs of numbers that multiply to form a given number	Fractions: The Meaning of Fractions Fractions: Addition & Subtraction
finance charge	the cost of borrowing money	Percent Applications
fraction	a way of showing parts of a whole; made up of two numbers, the numerator and the denominator	Fractions: The Meaning of Fractions
fuel gauge	the dial on a car's dashboard that indicates how much gasoline is in the car's tank	Fractions: Addition & Subtraction

G

Word	Definition	Student Book
gallon	a customary measurement for liquid 1 gallon = 3.8 liters	Fractions: The Meaning of Fractions Fractions: Addition & Subtraction
general admission	the cost of attending an event (concert, play, etc.)	Whole Numbers: Addition & Subtraction
greatest common factor	the largest number that two numbers can be divided by	Fractions: The Meaning of Fractions Fractions: Addition & Subtraction Fractions: Multiplication & Division

H

Word	Definition	Student Book
hour	a measure of time 1 hour = 60 minutes	Fractions: The Meaning of Fractions Fractions: Addition & Subtraction

I

Word	Definition	Student Book
improper fraction	a fraction where the numerator is greater than the denominator	Fractions: The Meaning of Fractions Fractions: Addition & Subtraction Fractions: Multiplication & Division The Meaning of Percent
income	the amount of money a person or business makes	Fractions: Multiplication & Division
increased	to get larger or expand	Decimals: Addition & Subtraction
ingredients	the parts of a recipe	Fractions: Multiplication & Division Ratio & Proportion
interest	the charge for loaned money	Percent Applications

GLOSSARY

Word	Definition	Student Book
L		
late fee	the money someone pays for returning something past the due date	Decimals: Addition & Subtraction
least common multiple	the smallest number that two numbers can be divided into	Fractions: Addition & Subtraction
like fractions	fractions that have the same denominator	Fractions: Addition & Subtraction
line segment	a straight line between two points	Fractions: Addition & Subtraction Fractions: Multiplication & Division
line up	to make sure the decimal points are in a line	Decimals: Addition & Subtraction
lowest common denominator	the smallest multiple of the denominators of fractions	Fractions: Addition & Subtraction
M		
mileage	total distance between two points, measured in miles	Decimals: Multiplication & Division
mile	a customary measurement of distance 1 mile = 1.6 kilometers	Fractions: The Meaning of Fractions Fractions: Addition & Subtraction
minus	to subtract	Whole Numbers: Addition & Subtraction
mixed decimal	the combination of a whole number and a decimal	Decimals: Addition & Subtraction
mixed number	the combination of a whole number and a fraction	Fractions: The Meaning of Fractions Fractions: Addition & Subtraction Fractions: Multiplication & Division The Meaning of Percent
monthly payment	money paid each month	Decimals: Multiplication & Division
multiple	a number into which another number may be divided with a remainder of zero	Whole Numbers: Multiplication & Division Fractions: Addition & Subtraction
multiplication	the process of adding a number to itself a certain number of times	Whole Numbers: Multiplication & Division
N		
number relation symbol	symbols that explain the relationship between two numbers: less than < greater than > is equal to = is not equal to ≠	Whole Numbers: Addition & Subtraction Whole Numbers: Multiplication & Division Decimals: Addition & Subtraction Decimals: Multiplication & Division Fractions: Addition & Subtraction Fractions: Multiplication & Division Ratio & Proportion The Meaning of Percent

Word	Definition	Student Book
number sentence	a complete math problem	Whole Numbers: Addition & Subtraction
numerator	the top part of a fraction	Fractions: The Meaning of Fractions Fractions: Addition & Subtraction Fractions: Multiplication & Division

O

Word	Definition	Student Book
operation symbol	the symbol (+, –, x, ÷) that tells you what to do with a math problem—a written sign used to represent an operation	Whole Numbers: Addition & Subtraction Whole Numbers: Multiplication & Division Decimals: Addition & Subtraction Decimals: Multiplication & Division Fractions: Multiplication & Division
opposite	to do the reverse (as in add and subtract)	Whole Numbers: Addition & Subtraction
options	choices	Whole Numbers: Addition & Subtraction
order	the specific arrangement of separate numbers or operations	Ratio & Proportion
overtime	to work more than the required time	Decimals: Multiplication & Division Fractions: Addition & Subtraction
oz (ounce)	a customary measurement used for liquid 8 ounces = 1 cup	Fractions: Multiplication & Division

P

Word	Definition	Student Book
pattern	a consistent series of numbers	Whole Numbers: Addition & Subtraction
per	for each	Decimals: Multiplication & Division Ratio & Proportion
percent	a way of expressing a number as the part of a whole; the word *percent* means *out of 100*	The Meaning of Percent Percent Applications
placeholder	a number (usually 0) that keeps a place in a problem and does not change the value of a number	Decimals: Multiplication & Division
place value	the name given to the space where a number is written	Whole Numbers: Addition & Subtraction Whole Numbers: Multiplication & Division Decimals: Addition & Subtraction
plus	to add	Whole Numbers: Addition & Subtraction
portion	a part of the whole	Fractions: Addition & Subtraction
pound	a customary measurement of weight 2.2 pounds = 1 kilogram	Fractions: The Meaning of Fractions Fractions: Addition & Subtraction
premium	the amount paid for an insurance policy	Percent Applications
principal	a sum of money, usually in a bank or investment account	Percent Applications
product	the answer to a multiplication problem	Fractions: The Meaning of Fractions Fractions: Multiplication & Division

Word	Definition	Student Book
proper fraction	a fraction where the numerator is smaller than the denominator	Fractions: The Meaning of Fractions
proportion	an expression made up of two equal ratios	Ratio & Proportion

Q

Word	Definition	Student Book
quantity	the amount of something	Whole Numbers: Multiplication & Division
quoted	a stated price	Whole Numbers: Addition & Subtraction
quotient	the answer to a division problem	Whole Numbers: Multiplication & Division Decimals: Multiplication & Division

R

Word	Definition	Student Book
rate	a quantity: speed, interest paid, taxes	Fractions: Multiplication & Division
ratio	a comparison of two numbers	Ratio & Proportion The Meaning of Percent
reciprocal	when a fraction is inverted so the numerator becomes the denominator	Fractions: Multiplication & Division
recreation	play activities	Fractions: Multiplication & Division
reduction	to make smaller	Percent Applications
registered	to sign up for something such as a class	Whole Numbers: Addition & Subtraction
regroup (borrow)	to shift a number to a lower place value	Whole Numbers: Addition & Subtraction Decimals: Addition & Subtraction
regroup (carry)	to shift a number to a higher place value	Whole Numbers: Addition & Subtraction Whole Numbers: Multiplication & Division Decimals: Addition & Subtraction Decimals: Multiplication & Division
regular price	the cost of an item not on sale	Whole Numbers: Multiplication & Division Decimals: Addition & Subtraction Whole Numbers: Addition & Subtraction
remainder	the number that is left over when a division problem doesn't divide evenly	Whole Numbers: Multiplication & Division Fractions: The Meaning of Fractions The Meaning of Percent Percent Applications
remaining balance	the amount left in an account	Decimals: Multiplication & Division
rename	to change a mixed or whole number into a fraction	Fractions: Addition & Subtraction Fractions: Multiplication & Division The Meaning of Percent Percent Applications

Word	Definition	Student Book
reserved seats	seats that are being held for someone	Whole Numbers: Addition & Subtraction
round	an estimation strategy where a number is increased or decreased to a given place value, usually ending in 0	Whole Numbers: Multiplication & Division

S

Word	Definition	Student Book
salary	the amount of money a person is paid in one year	Percent Applications
sale price	the reduced cost of an item	Whole Numbers: Multiplication & Division Decimals: Addition & Subtraction Whole Numbers: Addition & Subtraction
savings account	a bank account used to save money	Decimals: Addition & Subtraction Fractions: Multiplication & Division Percent Applications
shift	the time period a person works	Whole Numbers: Multiplication & Division Fractions: Multiplication & Division
simplify (reduce)	to make the numbers in a fraction smaller without changing the value of the fraction	Fractions: The Meaning of Fractions Fractions: Addition & Subtraction Fractions: Multiplication & Division Ratio & Proportion The Meaning of Percent Percent Applications
standard time (zone)	the time of day during the months not included in daylight savings time (eastern standard time, central standard time, mountain standard time, pacific standard time)	Whole Numbers: Multiplication & Division
subscription	an order for magazines	Whole Numbers: Multiplication & Division
subtraction	to find the difference between numbers	Whole Numbers: Addition & Subtraction
sum	the answer to an addition problem	Whole Numbers: Addition & Subtraction Fractions: The Meaning of Fractions
survey	information gathered about a given topic	Fractions: Multiplication & Division
symbol	(see operation symbol)	

T

Word	Definition	Student Book
take away	to subtract	Whole Numbers: Addition & Subtraction
tax	money paid to the government	Decimals: Addition & Subtraction
tax rate	the percentage of money paid to the government	Percent Applications
tbsp (tablespoon)	a customary unit of measurement 4 tbsp = $\frac{1}{4}$ cup	Fractions: Multiplication & Division

GLOSSARY

Word	Definition	Student Book
temperature	the measure of hot and cold	Decimals: Multiplication & Division
term	one of the numbers that makes up a ratio or fraction	Ratio & Proportion
times	when two numbers are multiplied	Whole Numbers: Multiplication & Division
tip	money left for a waiter or waitress	Decimals: Addition & Subtraction The Meaning of Percent Percent Applications
total earnings	the amount of money a person makes	Decimals: Multiplication & Division
tsp (teaspoon)	a customary unit of measurement 3 tsp = 1 tbsp	Fractions: Multiplication & Division

U

Word	Definition	Student Book
unit price	the cost of one item or one unit	Whole Numbers: Multiplication & Division Decimals: Multiplication & Division Ratio & Proportion
unit rate	the rate for one unit of a given quantity	Ratio & Proportion
units of measure	any standard value for measuring quantity	Decimals: Multiplication & Division

W

Word	Definition	Student Book
weekly allowance	money given to a person to spend each week	Percent Applications
whole number	a number beginning with 0, 1, 2, 3, 4, 5, 6, 7, 8, or 9	Decimals: Multiplication & Division Fractions: The Meaning of Fractions Fractions: Addition & Subtraction Fractions: Multiplication & Division